BIRKHÄUSER

Modern Birkhäuser Classics

Many of the original research and survey monographs in pure and applied mathematics published by Birkhäuser in recent decades have been groundbreaking and have come to be regarded as foundational to the subject. Through the MBC Series, a select number of these modern classics, entirely uncorrected, are being re-released in paperback (and as eBooks) to ensure that these treasures remain accessible to new generations of students, scholars, and researchers.

Alexander Lubotzky

Discrete Groups, Expanding Graphs and Invariant Measures

Appendix by Jonathan D. Rogawski

Ferran Sunyer i Balaguer
Award winning monograph

Reprint of the 1994 Edition

Birkhäuser Verlag
Basel · Boston · Berlin

Author:

Alexander Lubotzky
Einstein Institute of Mathematics
Hebrew University
Jerusalem 91904
Israel
e-mail: alexlub@math.huji.ac.il

Originally published under the same title as volume 125 in the Progress in Mathematics series by Birkhäuser Verlag, Switzerland, ISBN 978-3-7643-5075-8

1991 Mathematics Subject Classification 22E40, 43A05, 11F06, 11F70, 05C25, 28C10

Library of Congress Control Number: 2009937803

Bibliographic information published by Die Deutsche Bibliothek
Die Deutsche Bibliothek lists this publication in the Deutsche Nationalbibliografie; detailed bibliographic data is available in the Internet at <http://dnb.ddb.de>.

ISBN 978-3-0346-0331-7 Birkhäuser Verlag AG, Basel · Boston · Berlin

Basel · Boston · Berlin
P.O. Box 133, CH-4010 Basel, Switzerland
Part of Springer Science+Business Media
Printed on acid-free paper produced of chlorine-free pulp. TCF ∞

ISBN 978-3-0346-0331-7 e-ISBN 978-3-0346-0332-4

9 8 7 6 5 4 3 2 1 www.birkhauser.ch

Table of Contents

לזכר סבי

אשר ב"ר יצחק הכהן בליזובסקי

0 Introduction

In the last fifteen years two seemingly unrelated problems, one in computer science and the other in measure theory, were solved by amazingly similar techniques from representation theory and from analytic number theory. One problem is the explicit construction of expanding graphs («expanders»). These are highly connected sparse graphs whose existence can be easily demonstrated but whose explicit construction turns out to be a difficult task. Since expanders serve as basic building blocks for various distributed networks, an explicit construction is highly desirable. The other problem is one posed by Ruziewicz about seventy years ago and studied by Banach [Ba]. It asks whether the Lebesgue measure is the only finitely additive measure of total measure one, defined on the Lebesgue subsets of the n-dimensional sphere and invariant under all rotations.

The two problems seem, at first glance, totally unrelated. It is therefore somewhat surprising that both problems were solved using similar methods: initially, Kazhdan's property (T) from representation theory of semi-simple Lie groups was applied in both cases to achieve partial results, and later on, both problems were solved using the (proved) Ramanujan conjecture from the theory of automorphic forms. The fact that representation theory and automorphic forms have anything to do with these problems is a surprise and a hint as well that the two questions are strongly related.

The main goal of these notes is to present the two problems and their solutions from a unified point of view. We will explore how both solutions are just two different aspects of the same phenomenon: for some group G, the trivial one-dimensional representation is isolated from some subclass of irreducible unitary representations of G. Kazhdan's property (T) is precisely a property of the above. The Ramanujan conjecture also has such an interpretation, from which the two solutions can be deduced. In fact, the highlight of these notes is a single arithmetic group Γ embedded naturally in a direct product $G = G_1 \times G_2$ where G_1 is a real Lie group and G_2 a p-adic Lie group. The Ramanujan conjecture, in its representation theoretic formulation, controls the G-irreducible representations appearing in $L^2(\Gamma\backslash G)$. Using this result for the projection of Γ into $G_1 = SO(3)$ yields an affirmative solution to the Banach-Ruziewicz problem (for the sphere S^2 – but also more – see Chapter 7). Projecting Γ to the second factor $G_2 = PGL_2(\mathbb{Q}_p)$ and applying the same result, one obtains expanding graphs (as quotient graphs of the tree associated with $PGL_2(\mathbb{Q}_p)$ modulo the action of congruence subgroups of Γ).

There are very few new results in these notes. The main intention is to reproduce the existing solutions of the two problems in a way which stresses their unity. We also elaborate on the connection between the above two problems and other topics, e.g., eigenvalues of the Laplacian of Riemannian manifolds, Selberg's Theorem $\lambda_1 \geq \frac{3}{16}$ for arithmetic hyperbolic surfaces, the combinatorics of some finite simple groups, numerical analysis on the sphere and more.

Here is a brief chapter-by-chapter description of the contents; a more detailed one may be found at the beginning of each chapter.

In Chapter 1 we survey and illustrate the importance of expanders and prove their existence via counting arguments. In the second chapter we present the Hausdorff-Banach-Tarski paradox, which is a motivation for the Ruziewicz problem (and also plays a role in its solution) and show that the Ruziewicz problem for $n = 1$ has a negative answer. In Chapter 3, after introducing property (T) for Lie groups and their lattices, we apply it to solve affirmatively the Ruziewicz problem for $n \geq 4$. In addition, we use it to give an explicit construction of expanders which, however, are not as good as the ones whose existence is established by counting arguments. In the next chapter we connect our material with eigenvalues of the Laplacian operator of Riemannian manifolds and of graphs. We also bring to the fore Selberg's Theorem, which can be used to give interesting explicit expanders. The problem of expanders is translated into a problem about eigenvalues, and the important notion of Ramanujan graphs is introduced.

Chapter 5 is an introduction to the representation theory of PGL_2 over the reals and the p-adics. This is a fairly well-known topic, but we stress the unified treatment of both cases as well as the close connection with the eigenvalues of the corresponding Laplacians. We also present the tree associated with $PGL_2(\mathbb{Q}_p)$ analogous to the hyperbolic upper half-plane associated with $PGL_2(\mathbb{R})$. The material of this chapter is a necessary background for the sequel. We continue in Chapter 6, where we quote the works of Deligne and Jacquet-Langlands. The next chapter merges all the ingredients to completely solve Ruziewicz's problem, as well as to construct Ramanujan graphs. We also present there some of the remarkable properties of these graphs.

The next two chapters, Chapters 8 and 9, bring some miscellaneous topics related to the above: Chapter 8 contains applications to finite simple groups as well as some other methods to construct Ramanujan graphs (but of unbounded degree) and Ramanujan diagrams, while Chapter 9 brings a pseudo-random method to distribute points on the sphere, which is an application of the above methods and uses the same group Γ from Chapter 7. This method is of importance for numerical analysis on the sphere and elsewhere. Only the cases of S^2 and S^3 have satisfactory results, while the higher-dimensional cases (which can be thought as «quantified» Ruziewicz problems) are still open. This problem and many more are described in Chapter 10.

The Appendix, by Jonathan Rogawski, gives more details and elaborates on the material described in Chapters 5 and 6. While in the body of the book we emphasise the representation theoretic formulation of the Ramanujan (-Petersson) conjecture, the Appendix explains its classical form and the connection to modular forms. Rogawski also explains the Jacquet–Langlands theory and gives indications of how that conjecture was proved by Deligne and how the Jacquet–Langlands theory enables us to apply it for quaternion algebras.

These notes discuss essentially the same problems as those in the book of P. Sarnak [Sa2]. Still, the perspectives are very different, and this influenced our decision to keep the books separate. The reader is highly encouraged, however, to consult that book, as well as the two excellent exposition papers on these subjects by F. Bien [Bi] and Y. Colin de Verdiere [CV2]. This book is not self-contained, but we hope it is accessible to second year graduate students. The choice of what to prove and what just to survey was very subjective. Generally speaking, I tried to write it in a form of something I wish had existed when, eight years ago, I made my first steps into these subjects without specific background in any of them. I hope these notes will pave a more convenient road for those to follow into this exciting cross-section of analytic number theory, Lie groups, combinatorics, measure theory and differential geometry.

This monograph started out as notes prepared for the CBMS-NSF conference under the same title held at the University of Oklahoma during May 1989. It is based in part on notes from a Hebrew University seminar held in 1985/6, and courses given by the author at Yale University in 1988/9 and Columbia University in the fall of 1989. We are grateful to the participants of the seminars and courses for valuable discussions, remarks and encouragement. Special thanks to E. Shamir, H. Furstenburg, G. Kalai, N. Alon, S. Gelbart, M. Magidor, G.D. Mostow, W. Feit, R. Howe, I. Piatetski-Shapiro, Z. Rudnick, S. Mozes, Y. Peres, J. Rogawski, H. Bass and J.Dodziuk.

Many thanks to W. M. Kantor, A. Magid, S. Mozes, S. Adams, P. de la Harpe, A. Valette, M. Picardello, F. Williams, B. Mohar, D. Rockmore and B. Weiss for valuable advice and criticism on some parts of these notes.

Above all, I have the pleasure of thanking Andy Magid, Moshe Morgenstern and Peter Sarnak. From Peter I learnt much of the relevant background, Andy with his enthusiasm encouraged me to embark on this long voyage, while Moshe is responsible for my reaching a safe haven.

Jerusalem, December 1993 A.L.

1 Expanding Graphs

1.0 Introduction

The central theme of this chapter and of much of this book is expanders. Expanders are sparse graphs with strong connectivity properties. More precisely, a k-regular graph X with n vertices is called a c-expander $(0 < c \in \mathbb{R})$ if for every subset A of X, $|\partial A| \geq c(1 - \frac{|A|}{|X|})|A|$ where $\partial A = \{y \in X \mid d(y, A) = 1\}$ is the boundary of A. A family of expanders is a family of k-regular graphs (for a fixed k and n going to infinity) which are all c-expanders for the same c.

Expander graphs play an important role in computer science, especially as basic building blocks for various communication networks (but not only). It is not difficult to show by counting arguments that families of expanders do exist (1.2.1). But for applications, an explicit construction is desirable. This is a much more difficult problem which will be addressed in later chapters.

1.1 Expanders and their applications

This section will be devoted to the definition and basic properties of expanders. We will work only with k-regular graphs (i.e., graphs with the same valency at each vertex). This is because they are anyway those which appear in all our examples and applications. The definitions and basic results here can be easily extended to general graphs (at least for those with bounded degree).

Definition 1.1.1. *A finite regular graph $X = X(V, E)$ with a set V of n vertices and of degree k (so the set of edges E contains $\frac{kn}{2}$ edges) is called an (n, k, c)-expander if for every subset A of V,*

$$|\partial A| \geq c\left(1 - \frac{|A|}{n}\right)|A| \tag{1.1.1}$$

where $\partial A = \{y \in V \mid d(y, A) = 1\}$ is the «boundary» of A and d is the distance function on X.

Remarks 1.1.2. (i) Every finite connected k-regular graph is an expander for some $c > 0$ in a trivial way. The notion is of interest only when one considers an infinite family of graphs. In applications, one usually wants a family of (n, k, c)-expanders where n is going to infinity and k and c are fixed. Usually (but not always) one prefers k to be as small as possible, and it is always desirable that c will be as large as possible.

(ii) There are also some other definitions of expanders in the literature. All are essentially equivalent up to some change in the constants. The basic idea is always that every set in some class of subsets is guaranteed to expand by some fixed amount. A quite common definition is the following:

Let $c' > 0$. An (n, k, c')-expander X is a bi-partite k-regular graph with two parts I and O where the edges go from I to O, $|I| = |O| = n$ and for any $A \subset I$ with $|A| \leq \frac{n}{2}$ we have $|\partial A| \geq (1 + c')|A|$.

Given an expander $X = X(V, E)$ in the sense of (1.1.1), we may get an expander in the sense of (1.1.2(ii)) by taking its double cover, i.e., taking I and O to be two disjoint copies of V and connecting an input in I to its «twin» in O and to the twins of all its neighbors in X. Conversely, given an expander in the sense of (1.1.2(ii)), one gets an expander in the sense of (1.1.1) by identifying I and O (using P. Hall Marriage Lemma). In both directions we might have to change the constants c and c'.

In analogue to the Cheeger constant of a Riemannian manifold (see [Ch]), it is natural to define:

Definition 1.1.3. *Let $X = X(V, E)$ be a finite graph. Define the Cheeger constant of X, denoted $h(X)$, by:*

$$h(X) = \inf_{A,B \subseteq V} \frac{|E(A, B)|}{\min(|A|, |B|)}$$

where the infimum runs over all the possibilities of a disjoint partition $V = A \cup B$ and $E(A, B)$ is the set of edges connecting vertices in A to vertices in B.

The following Proposition follows from the definitions. The proof is left to the reader.

Proposition 1.1.4. *Let X be a k-regular graph with n vertices. Then*

(i) *If X is an (n, k, c)-expander, then $h(X) \geq \frac{c}{2}$*

(ii) *X is an $(n, k, \frac{h(X)}{k})$-expander.* □

So, talking about a <u>fixed</u> k and varied n, to get a family of (n, k, c)-expanders for *some* $c > 0$, is equivalent to having a family of k-regular graphs whose Cheeger constant is bounded away from 0.

In describing the broad spectrum of applications of expanders, we cannot do better than quoting M. Klawe's introduction of [Kl2]:

> «*The study of the complexity of graphs with special connectivity properties originated in switching theory, motivated by problems of designing networks able to connect many disjoint sets of users, while only using a small number of switches. An example of this type of graph is a superconcentrator, which is an acyclic directed graph with n inputs and n outputs such that given any pair of subsets A and B of the same size,*

of inputs and outputs respectively, there exists a set of disjoint paths joining the inputs in A to the outputs in B. Some other examples are concentrators, nonblocking connectors and generalized connectors (see [C1], [P2] for more details). There is a large body of work searching for optimal constructions of these graphs (Pinsker [Pi], Bassalygo and Pinsker [BP], Cantor [Ca], Ofman [Of], Masson and Jordan [MJ], Pippenger [P1], [P2], Chung [C1]). So far all optimal explicit constructions depend on expanding graphs of some sort.

«Superconcentrators have also proved to be useful in theoretical computer science. By showing that the computation graphs of straight line programs for problems such as polynomial multiplication, the Fourier transform and matrix inversion must be superconcentrators, it has been possible to establish nonlinear lower time bounds and time-space trade-offs for these problems assuming certain models of computation (Valiant [Va], Abelson [Ab], Ja'Ja' [Ja], Tompa [To]).

«These space-time tradeoffs are obtained via a game known as pebbling, which is played on acyclic directed graphs and mimics the storage of temporary results during a straight-line computation. In considering the problem of pebbling an arbitrary acyclic directed graph, expanding graphs have been used in several instances to construct graphs which are (in some sense) hardest to pebble, hence establishing lower bounds in space-time trade-offs (Lengauer and Tarjan [LT], Paul and Tarjan [PT], Paul, Tarjan and Celoni [PTC], Pippenger [P3]).

«Expanding graphs have also been used to construct sparse graphs with dense long paths (Erdos, Graham and Szemeredi [EGS]). Interest in sparse graphs with dense long paths stem from studying the complexity of Boolean functions, and more recently from problems of designing fault-tolerant microelectronic chips. Paul and Reischuk strengthened this result by constructing (still using expanding graphs) sparse graphs of bounded in-degree with dense long paths, which is of interest since computation graphs have bounded in-degree.

«Perhaps the most practical applications of expanding graphs occur in the two most recent results. Ajtai, Komlos and Szemeredi [AKS] have announced the construction of an oblivious sorting network using $O(n \log n)$ *comparators, and having depth* $O(\log n)$*. Again, expanding graphs form the basic components, and of course, the explicit construction of the sorting network depends on the explicit construction of expanding graphs. The problem of constructing such a sorting network has been open for twenty years, which perhaps illustrates best the unexpected power of expanding graphs. Finally, expanding graphs have been used by Karp and Pippenger [KP] to design an algorithm which can be applied to virtually all the well-known Monte-Carlo algorithms to reduce the number of uses of a randomization resource (i.e., coin-flips or*

calls to a random number generator) while still maintaining polynomial running time.»

To the above quite impressive list, we can add some more recent papers on applications to sorting in rounds (Alon [A2], Pippenger [P4]), routing networks (Leighton-Maggs [LM], Upfal [Up]) and other papers such as [FFP], [BGG] and the references therein.

In a completely different direction, Friedman and Pippenger [FP] proved the following interesting result:

Theorem 1.1.5. *If X is a non-empty graph such that for every $A \subseteq V(X)$ with $|A| \leq 2\ell - 2$, $|\partial A| \geq d|A|$, then X contains every tree with ℓ vertices and maximum degree at most d.*

For the readers who are not familiar with the above-mentioned topics, we will sketch in more detail one example, which is probably the best known application. This is the superconcentrator. (Here we follow [P1], [GG], [C1] and [Bs]).

Definition 1.1.6. *An (n, ℓ)-superconcentrator is a directed acyclic graph with n inputs, and n outputs, having at most ℓn edges, such that for any choice of r inputs and r outputs there is a collection of r disjoint paths starting with the input set and ending in the output set. ℓ is called the density of the superconcentrator.*

To construct a superconcentrator, one first builds a bounded concentrator.

Definition 1.1.7. *For $\alpha \leq \theta < 1$, an (n, θ, k, α)-bounded concentrator is a bipartite graph X with n inputs, θn outputs and at most kn edges, such that every input subset A with $|A| \leq \alpha n$ maps to an output set $\partial(A)$ at least as large. (Note that by P. Hall Marriage Lemma [Or] there is also a perfect matching between A and a subset of its neighbors).*

Superconcentrators are obtained from bounded concentrators in a recursive construction which goes like this: Connect input i directly to output i for $1 \leq i \leq n$ using n edges. Given an input subset and an equinumerous output subset, use these direct lines if possible to connect inputs to outputs by paths of length one. After that, there remain fewer than $n/2$ unmatched inputs to be connected to a set of unmatched outputs. Use an $(n, \theta, k, \frac{1}{2})$-bounded concentrator to concentrate the unmatched inputs into a new «input space» of size θn. Similarly concentrate the unmatched outputs into an «output space» of size θn. Then use a size θn superconcentrator to match the unmatched. Start the recursive definition by using, say, a complete n^2 edge graph as a superconcentrator when n is small.

$(n, \theta, k, \frac{1}{2})$-bounded concentrators are obtained from expanders as follows: Let r be a positive integer and assume we have an (m, k, c)-expander where $c \geq \frac{2r^2}{(r-1)(r^2+1)}$. Then we build an $(n = m(r+1)/r,\ \theta = r/(r+1),\ k' = (k+1)r/(r+1),\ \alpha = \frac{1}{2})$-bounded concentrator by breaking first the input set into a «big part» of size m and a «small part» of size m/r. Use a (m, k, c)-(bi-partite) expander to connect the big part of the input set to the output set (which has size m). Divide the output set into m/r clumps of size r and connect each of the inputs in the small set to all of the members of one of the output clumps, each input to a different clump. This is indeed a bounded concentrator: Let s be the number of inputs in a given subset with $s \leq m(r+1)/2r$. We need to show that these connect to a total of at least s outputs. If at least s/r of these appear in the small set, then we are done because $r \cdot s/r \geq s$. Otherwise, there will be at least $q = s(r-1)/r$ of these in the large input set. Since these feed the expander, we will succeed if the expansion factor is at least $r/(r-1)$; namely, if $1 + c(1 - \frac{q}{m}) \geq \frac{r}{r-1}$. But $1+c(1-\frac{q}{m}) = 1+c(1-\frac{s(r-1)}{rm}) \geq 1+c(1-\frac{r^2-1}{2r^2})$, and the last is large enough when $c \geq \frac{2r^2}{(r-1)(r^2+1)}$.

Working all the constants in the constructions, one gets:

Theorem 1.1.8. (Gaber-Galil [GG]) *Assuming we can construct for every n an $(n, k, 1/(r-1))$-expander, then we can construct a family of $(n, \ell = (2k+3)r+1)$-superconcentrators for infinitely many values of n.*

In the next section we will discuss the question of existence of expanders. Their existence will be proved easily, but for «real world» applications such as superconcentrators, one needs explicit constructions. This is a much harder task to be discussed in later chapters.

1.2 Existence of expanders

The existence of expanders is a relatively easy question. By counting arguments below we will see that (at least when k and n are large enough) «most» k-regular graphs are expanders. When one wants to check that a family of k-regular graphs is a family of expanders (with <u>some</u> $c > 0$), it is enough to check condition (1.1.1) only for sets of size $\leq \frac{n}{2}$ (since k is fixed). This is of no real help in practice since there are still too many subsets of size $\leq \frac{n}{2}$. It helps, however, a little bit with the counting arguments:

Proposition 1.2.1. *Let $k \geq 5$ be an integer and $c = \frac{1}{2}$. Then some k-regular graphs on n vertices satisfy $|\partial(A)| \geq c|A|$ for subsets A of size $\leq \frac{n}{2}$. For n large, most k-regular graphs on n vertices satisfy this. In particular, they are expanders (see Remark 1.1.2 (ii)).*

Remarks. (i) We will bring a simple proof following [Sa2]. More precise results can be obtained; see for example Pinsker [Pi], Pippenger [P1] and Chung [C1].

(ii) The proposition says that most k-regular graphs are expanders. There is, however, some ambiguity in this statement around the term «most». Here one should specify the way he counts the k-regular graphs. The most natural way would be «up to isomorphism», but there is no good model for that. A useful model for random k-regular graphs was proposed by Bollobás (see [B4]). But we will not go into details here. The way we prove our Proposition is sufficiently convincing that indeed most k-regular graphs are expanders in any reasonable sense of the word «most».

Proof. It will be most convenient to prove it for bi-partite expanders as defined in Remark 1.1.2 (ii).

Let $I = O = \{1, 2, \ldots, n\}$ and construct the bi-partite graph X by taking permutations $\pi_1, \ldots, \pi_k$ of $\{1, \ldots, n\}$ and joining each $i \in I$ with $\pi_1(i), \ldots, \pi_k(i) \in O$. This yields a bi-partite k regular graph. (In fact, it follows from König's Theorem (cf. [Or]) that every k-regular bi-partite graph is obtained this way.) What we claim is that for most choices of $\pi = (\pi_1, \ldots, \pi_k)$ the graph so obtained is a c-expander for $c = \frac{1}{2}$. There are $(n!)^k$ such k-tuples. A k-tuple $\pi = (\pi_1, \ldots, \pi_k)$ is called bad if the associated graph is not an $\frac{1}{2}$-expander, i.e., if for some $A \subset I$ with $|A| \le \frac{n}{2}$ there is a subset $B \subset O$ with $|B| = [\frac{3}{2}|A|]$ for which $\pi_j(A) \subset B$ for every $j = 1, \ldots, k$.

Let A be given of order t and B of order $m = \frac{3}{2}t$. The number of bad π's corresponding to A and B is $(m(m-1)\cdots(m-t+1)(n-t)!)^k = (\frac{m!(n-t)!}{(m-t)!})^k$. Hence the total number of bad choices, say β, is bounded by: $\beta \le \sum_{t \le \frac{n}{2}} \binom{n}{t}\binom{n}{m}(\frac{m!(n-t)!}{(m-t)!})^k = \sum_{t \le \frac{n}{2}} R(t)$. By checking $R(t)/R(t+1)$ we see that for $t < \frac{n}{3}$ this is at least 1, and so at this range $R(t)$ gets its maximum for $t = 1$. For $\frac{n}{3} \le t \le \frac{n}{2}$, $\binom{n}{t}\binom{n}{m} \le 2^{2n}$ and $\frac{m!(n-t)!}{(m-t)!}$ gets its maximum either at $t = \frac{n}{3}$ or at $t = \frac{n}{2}$. Now, it is easy to check that $(n/2)[R(1) + R(\frac{n}{3}) + R(\frac{n}{2})]/(n!)^k$ tends to 0 when n is going to ∞.$\square$

Though existence is easy, explicit construction turns out to be very difficult. There are up to date only two methods of explicit constructions of families of expanders: The first was Margulis [M1] who constructed expanders with the help of Kazhdan property (T) from representation theory of semi-simple Lie groups. (See Chapter 3, and for some variants of his construction see [GG], [JM] and [AM]). Secondly, Lubotzky-Phillips-Sarnak [LPS1], [LPS2] and Margulis [M6] used Ramanujan conjecture (as proved by Eichler [Ei] and Deligne [Dl1], [Dl2]) for that purpose. (A variant of it, which uses Selberg's Theorem instead, is presented in Chapter 4; see also [Bu 3] and [Br 1].) See more in Chapters 4 and 7.

2 The Banach-Ruziewicz Problem

2.0 Introduction

The second central problem of this book is presented in this chapter. This is the Banach-Ruziewicz problem (known also as the Ruziewicz problem), which asks whether the Lebesgue measure λ is the only finitely additive measure defined on the Lebesgue subsets of the n-dimensional sphere S^n, invariant under the group of rotations $O(n+1)$ and with total measure one.

Assuming «countable additive» instead of finitely additive implies the uniqueness (2.2.9). The interest in finitely additive invariant measures was motivated by the Hausdorff-Banach-Tarski paradox claiming that for $n \geq 2$, S^n can be decomposed into finitely many pieces, from which one can reconstruct two copies of S^n by using only rotations from $O(n+1)$. This is impossible for S^1, essentially because S^1 is an amenable group and hence there is an invariant mean on $L^\infty(S^1)$. Moreover, there are many invariant means on $L^\infty(S^1)$, which implies a negative solution to the Ruziewicz problem. The negative solution for $n = 1$ was proved by Banach in 1921. The problem for $n \geq 2$ was open up to recent years, when it was solved affirmatively using Kazhdan property (T) and Ramanujan conjecture. The affirmative part and its ramifications will occupy a significant part of these notes. Meanwhile in this chapter, we describe the background of the problem, including the Hausdorff-Banach-Tarski paradox. This paradox is not only a motivation, but it also plays a role in the affirmative solution for $n \geq 2$. We also present the negative solution in case $n = 1$.

2.1 The Hausdorff-Banach-Tarski paradox

Definition 2.1.1. *Let G be a group acting on a set X, and A, B subsets of X. A and B are said to be (G)-equidecomposable if A and B can each be partitioned into the same finite number of respectively G-congruent pieces. Formally, $A = \bigcup_{i=1}^{n} A_i$ and $B = \bigcup_{i=1}^{n} B_i$ where $A_i \cap A_j = B_i \cap B_j = \phi$ for $1 \leq i < j \leq n$ and there are $g_1, \ldots, g_n \in G$ such that for each $1 \leq i \leq n$, $g_i(A_i) = B_i$.*

We will denote it $A \sim B$ and write $A \underset{\sim}{<} B$ if $A \sim C$ for some subset C of B. A *realization* h of $A \sim B$ is a bijection $h : A \to B$ such that there exist decompositions as above $A = \bigcup_{i=1}^{n} A_i$, $B = \bigcup_{i=1}^{n} B_i$ and $g_1, \ldots, g_n \in G$ with $h(a_i) = g_i(a_i)$ for every $a_i \in A_i$ and every $i = 1, \ldots, n$. It is easy to see that if $h : A \to B$ is a realization of $A \sim B$ and $S \subseteq A$, then $S \sim h(S)$. We will freely use this fact.

Proposition 2.1.2. (Banach-Schröder-Bernstein Theorem) *Suppose G acts on X and $A, B \subseteq X$. If $A \lesssim B$ and $B \lesssim A$, then $A \sim B$.*

Proof. The fact that $A \lesssim B$ and $B \lesssim A$ implies the existence of bijections $f : A \to B_1$ and $g : A_1 \to B$ where $A \sim B_1 \subseteq B$ and $B \sim A_1 \subseteq A$. Let $C_0 = A \backslash A_1$ and by induction define C_{n+1} to be $g^{-1}f(C_n)$. Let $C = \bigcup_{n=1}^{\infty} C_n$. Then $g(A \backslash C) = B \backslash f(C)$. Thus $A \backslash C \sim B \backslash f(C)$. But we also have $C \sim f(C)$, so $A \sim B$. □

Corollary 2.1.3. *The following two conditions are equivalent:*

(i) *There are two proper disjoint subsets A and B of X such that $A \sim X \sim B$.*
(ii) *There are two proper disjoint subsets A and B such that $A \cup B = X$ and $A \sim X \sim B$.*

Proof. We have to show only (i) $\to$ (ii): But, $X \sim B \subseteq X \backslash A \subseteq X$, so by (2.1.2) $X \backslash A \sim X$, and we are done. □

Definition 2.1.4. *Let G be a group acting on a set X. X is said to be G-paradoxical if one (and hence both) of the conditions of (2.1.3) is satisfied. In simple words this says that X can be decomposed into finitely many pieces from which two copies of X can be rebuilt.*

Remark 2.1.5. Let A be a subset of X. Assume A is a pairwise disjoint union of subsets A_i, $A = \bigcup_{i=1}^{n} A_i$ and there exist $g_i \in G$, $i = 1, \ldots, n$ such that $\bigcup_{i=1}^{n} g_i(A_i) = X$. Then $A \sim X$.

Proof. By making the A_i's smaller we can ensure the $g_i(A_i)$ are disjoint, and therefore X is equidecomposable with some subset of A, and hence by (2.1.2) also with A. □

Example 2.1.6. (i) A free group F of rank 2 is F-paradoxical, where F acts on itself by left multiplication.

(ii) If F acts on a set X freely (i.e., every $1 \neq g \in F$ has no fixed points), then X is F-paradoxical.

Proof. (i) Say $F = F(a, b)$ and let A^+ (resp: A^-) be the set of all reduced words beginning with a (resp: a^{-1}). Similarly, define B^+ and B^-, and let $A = A^+ \cup A^-$ and $B = B^+ \cup B^-$. Now $A \sim F \sim B$ as $F = A^+ \cup aA^- = B^+ \cup bB^-$.

(ii) Let M be a set of representatives for the F-orbits in X (the axiom of choice is used here). For $T \in \{A^+, A^-, B^+, B^-\}$, let $X_T = T(M) = \{tm \mid t \in T,\ m \in M\}$. Then it is easy to check that $X_{A^+}, X_{A^-}, X_{B^+}, X_{B^-}$ are disjoint (recall F acts fixed point free!) and $X = X_{A^+} \cup aX_{A^-} = X_{B^+} \cup bX_{B^-}$. □

Proposition 2.1.7. *The group $SO(3)$ of orthogonal 3×3 real matrices of determinant 1 contains a free non-abelian group and in particular a free group on two generators.*

Remark. This proposition is well known and can be proved in many ways (see for example [Wa, Theorem 2.1] or [Ti]). The following proof is pretty special, but as we need this specific group later on, we bring it here, following [GVP, Chap. IX, §1] and [LPS4]. For the proof we will need the following classical Jacobi Theorem.

Theorem 2.1.8. *Let n be a positive integer. Then the number of solutions $(x_1, x_2, x_3, x_4) \in \mathbb{Z}^4$ with $x_1^2 + x_2^2 + x_3^2 + x_4^2 = n$ is $r_4(n) = 8 \sum_{\substack{d \mid n \\ 4 \nmid d}} d$. In particular, if $n = p$ is a prime then $r_4(p) = 8(p+1)$.*

An analytic proof of Theorem 2.1.8 can be found in [HW, Theorem 386].

We bring here an algebraic proof due to Hurwitz. We thank W. Feit for calling our attention to [Hu]. We give the proof only for the case $n = p$ is an odd prime, leaving the reader the task of deducing the general case from it (see also [HW, Chap. XX]).

Let $\tilde{H}(\mathbb{Z})$ be the ring of Hurwitz integral quaternions, i.e., $\{a_0 f + a_1 i + a_2 j + a_3 k \mid a_i \in \mathbb{Z}, i = 0, \dots, 3\}$ where $f = \frac{1}{2}(1 + i + j + k)$, $i^2 = j^2 = k^2 = -1$, $ij = -ji = k$, etc. (We prefer to work with the «maximal order» $\tilde{H}(\mathbb{Z})$ and not with $H(\mathbb{Z})$, as below, since $\tilde{H}(\mathbb{Z})$ is an Euclidean domain while $H(\mathbb{Z})$ is not – see [He, §7.4] or [Di].) $\tilde{H}(\mathbb{Z})$ is a principal ideal domain ([He, §7.4]) with a unit group U consisting of 24 elements. Two elements $\alpha, \beta \in \tilde{H}(\mathbb{Z})$ generate the same left ideal if and only if $\alpha = \varepsilon\beta$ for some $\varepsilon \in U$. We want to count the number of elements of norm p which is thus equal to 24 times the number of proper left ideals containing $p\tilde{H}(\mathbb{Z})$ properly. Now, since p is odd, $\tilde{H}(\mathbb{Z})/p\tilde{H}(\mathbb{Z}) = \tilde{H}(\mathbb{Z}/p\mathbb{Z}) \simeq M_2(\mathbb{F}_p)$, the 2×2 matrix algebra over the field $\mathbb{F}_p$. Thus, we should count the number of proper left ideals of $M_2(\mathbb{F}_p)$. These will be principal left ideals generated by the non-investible non-zero elements. Their number is: $p^4 - [(p^2 - 1)(p^2 - p)] - 1 = (p+1)(p^2 - 1)$. The group $\mathrm{GL}_2(\mathbb{F}_p)$ acts on them by left multiplication, and elements are on the same orbit iff they generate the same left ideal. These elements are of determinant 0 and are therefore conjugate to either $\begin{pmatrix} a & 0 \\ 0 & 0 \end{pmatrix}$ for some $0 \neq a \in \mathbb{F}_p$ or $\begin{pmatrix} 0 & 1 \\ 0 & 0 \end{pmatrix}$. In either case the stabilizer under the $\mathrm{GL}_2(\mathbb{F}_p)$-left multiplication is of order $p(p-1)$, and hence every orbit is of size $\#\mathrm{GL}_2(\mathbb{F}_p)/\ p(p-1) = p^2 - 1$. There are, therefore, $p+1$ left ideals and $24(p+1)$ elements of norm p in $\tilde{H}(\mathbb{Z})$. It is simple to check that $8(p+1)$ of them have integral entries while the rest do not. □

Now let p be a prime congruent to $1 \bmod 4$. Let $H(\mathbb{Z})$ denote the integral quaternions $H(\mathbb{Z}) = \{\alpha = a_0 + a_1 i + a_2 j + a_3 k \mid a_j \in \mathbb{Z}\}$, $i^2 = j^2 = k^2 = -1$,

$ij = -ji = k$, etc. Set $\bar{\alpha} = a_0 - a_1 i - a_2 j - a_3 k$ and let $N(\alpha)$ be the integer $\alpha\bar{\alpha} = \Sigma a_i^2$. The units of the ring $H(\mathbb{Z})$ are $\pm 1, \pm i, \pm j, \pm k$. Let S' be the set of all α with $N(\alpha) = p$. Since $p \equiv 1(4)$, only one of the a_i's will be odd. By (2.1.8), $|S'| = 8(p+1)$. The units act on this set, and each solution has exactly one associate $\varepsilon\alpha$, ε a unit, with $\varepsilon\alpha \equiv 1 \pmod 2$ and $a_0 > 0$. Let S be the set of these $p+1$ representatives. This set splits into distinct conjugate pairs $\{\alpha_1, \bar{\alpha}_1, \ldots, \alpha_s, \bar{\alpha}_s\}$ where $s = \frac{1}{2}(p+1)$. By a reduced word of length m in the elements of S we mean a word of length m in which no expression of the form $\alpha_j\bar{\alpha}_j$ or $\bar{\alpha}_j\alpha_j$ appears.

Lemma 2.1.9. ([GVP] or [LPS2]) *Every $\alpha \in H(\mathbb{Z})$ with $N(\alpha) = p^k$ can be expressed uniquely in the form $\alpha = \varepsilon p^r R_m(\alpha_1, \ldots, \bar{\alpha}_s)$ where ε is a unit, $2r+m = k$ and R_m is a reduced word in the α_i's of length m.*

Proof. To obtain the existence of such an expression we use the results of Hurwitz [Hu] and Dickson [Di] which show that for odd quaternions (i.e., those of odd norm) one has a theory of g.c.d. and the usual factorization (on the left and right) (see also [He]). Since α is odd and a quaternion is prime iff its norm is prime, we may write $\alpha = \gamma\beta$ with $N(\gamma) = p^{k-1}$ and $N(\beta) = p$. By the choice of S we can find a unit ε such that $\alpha = \gamma\varepsilon s_k$ with $s_k \in S$. Now repeat this for $\gamma\varepsilon$, etc. We eventually get $\alpha = \varepsilon s_1 \cdot \ldots \cdot s_k$ with $s_j \in S$. After performing cancellations (recall that $\alpha_i\bar{\alpha}_i = N(\alpha_i) = p$) we arrive at $\alpha = \varepsilon p^r R_m$ for some r and m. This proves the existence of such a decomposition.

We show the uniqueness by a counting argument (another way could be to use carefully the uniqueness of factorization of quaternions). First, the number of reduced words $R_m(\alpha_1, \ldots, \bar{\alpha}_s)$ is $(p+1)p^{m-1}$ for $m \geq 1$ and is 1 if $m = 0$. Hence the number of expressions $\varepsilon p^r R_m(\alpha_r, \ldots, \bar{\alpha}_s)$ with $2r + m = k$ is:

$$8\left(\sum_{0 \leq r < k/2} (p+1)p^{k-2r-1} + \delta(k)\right) = \sum_{j=0}^{k} p^j$$

where $\delta(k) = \begin{cases} 1 & \text{if } k \text{ is even} \\ 0 & \text{if } k \text{ is odd.} \end{cases}$. Hence the number of such expressions is:

$$8\left(\frac{p^{k+1}-1}{p-1}\right) = 8\sum_{d|p^k} d.$$

This is by (2.1.8) the number of $\alpha \in H(\mathbb{Z})$ with $N(\alpha) = p^k$. It follows that each such expression represents a distinct element. □

Corollary 2.1.10. *If $\alpha \equiv 1 \pmod 2$ and $N(\alpha) = p^k$, then $\alpha = \pm p^r R_m(\alpha_1, \ldots, \bar{\alpha}_s)$ with $2r + m = k$ and this representation is unique.* □

Consider now $H(\mathbb{Z}[\frac{1}{p}])$, i.e., all $\alpha = a_0 + a_1 i + a_2 j + a_3 k$ such that a_i ($i = 0, 1, 2, 3$) are p'-integral rational numbers, namely: only powers of p can appear in their denominators. Let $\Lambda'(2)$ be the set of all $\alpha \in H(\mathbb{Z}[\frac{1}{p}])$ such that $\alpha \equiv 1 \pmod 2$ and $N(\alpha) = p^\ell$ for some $\ell \in \mathbb{Z}$. $\Lambda'(2)$ is closed under multiplication, and if we identify α and β in $\Lambda'(2)$ whenever $\pm p^i \alpha = \beta$ for some $i \in \mathbb{Z}$, then the classes so obtained form a group with $[\alpha\beta] = [\alpha][\beta]$ and $[\alpha][\bar{\alpha}] = [1]$. Call this group $\Lambda(2)$.

Corollary 2.1.11. *$\Lambda(2)$ is a free group on the $s = \frac{p+1}{2}$ generators*

$$[\alpha_1], [\alpha_2], \dots, [\alpha_s].$$

Proof. Every $\alpha \in \Lambda'(2)$ is equivalent to one in $H(\mathbb{Z})$, and Corollary 2.1.10 ensures the unique representation of such as a word in $[\alpha_1], \dots, [\alpha_s]$, which exactly means that the group is free. □

We now look at the natural embedding

$$\sigma : H(\mathbb{Z}[\frac{1}{p}]) \to H(\mathbb{R}) \simeq \left\{ \begin{pmatrix} a+bi & c+di \\ -c+di & a-bi \end{pmatrix} \middle| a,b,c,d \in \mathbb{R} \right\}$$

$$\alpha = a + bi + cj + dk \xrightarrow{\sigma} \begin{pmatrix} a+bi & c+di \\ -c+di & a-bi \end{pmatrix}.$$

It is straightforward to check that σ induces a well-defined homomorphism $\tilde{\sigma} : \Lambda(2) \to H(\mathbb{R})^*/Z(H(\mathbb{R})^*)$, where $H(\mathbb{R})^*$ denotes the group of invertible real quaternions. This gives the promised embedding of a free group on s generators, and in particular of 2 generators, in $SO(3)$, since $SO(3) \simeq H(\mathbb{R})^*/Z(H(\mathbb{R})^*)$.

To see this recall that σ can be extended to give an isomorphism $H(\mathbb{C}) \simeq M_2(\mathbb{C})$, which is given explicitly by: $\alpha = a+bi+cj+dk \mapsto \begin{pmatrix} a+bi & c+di \\ -c+di & a-bi \end{pmatrix}$ and $\tilde{\sigma}$ is indeed an embedding into $SU(2)/\{\pm 1\} \simeq SO(3)$. Another way to see the isomorphism $H(\mathbb{R})^*/Z(H(\mathbb{R})^*) \simeq SO(3)$ is by considering the action by conjugation of $H(\mathbb{R})^*$ on the «imaginary quaternions», i.e., on $\{\alpha = bi + cj + dk \mid b,c,d \in \mathbb{R}\}$. This space with the quaternionic norm is isomorphic to the three-dimensional Euclidean space and the conjugation action preserves this norm. This defines a map from $H(\mathbb{R})^*$ into $O(3)$. As $H(\mathbb{R})^*$ is connected the image is in $SO(3)$, and the kernel is $Z(H(\mathbb{R})^*)$ as can be easily checked. Since $\dim(H(\mathbb{R})^*/Z(H(\mathbb{R})^*)) = \dim SO(3) = 3$, this is an isomorphism. Proposition 2.1.7 is now proved.

Corollary 2.1.12. *There is a countable subset D of S^2 – the two-dimensional sphere – such that $S^2 \backslash D$ is F-paradoxical for some subgroup F of $SO(3)$ (keeping $S^2 \backslash D$ invariant).*

Proof. Let F be a free subgroup of $SO(3)$ of rank 2. Every non-trivial element of $SO(3)$ has exactly two fixed points in S^2. Let $D = \{x \in S^2 | \exists 1 \neq r \in F, \gamma(x) = x\}$. Example 2.1.6(ii) implies that $S^2 \backslash D$ is F-paradoxical and therefore $SO(3)$-paradoxical. □

Proposition 2.1.13. *If D is a countable subset of S^2, then S^2 and $S^2 \backslash D$ are $SO(3)$-equidecomposable.*

Proof. Let ℓ be a line through the origin that misses the countable set D. Let θ be an angle such that for every integer $n \geq 0$, $\rho^n(D) \cap D = \phi$ where ρ^n is the rotation around ℓ of angle $n\theta$. Such θ does exist since D is countable. Let $\bar{D} = \bigcup_{n=0}^{\infty} \rho^n(D)$. Then $S^2 = \bar{D} \cup (S^2 \backslash \bar{D}) \sim \rho(\bar{D}) \cup (S^2 \backslash \bar{D}) = S^2 \backslash D$. □

Corollary 2.1.14. *S^2 is $SO(3)$-paradoxical.*

This last statement is (one form of) the Hausdorff-Banach-Tarski paradox.

Corollary 2.1.14 can be extended quite easily to S^n for every n, by induction on n: Assume $S^{n-1} = A \cup B$ is a paradoxical decomposition of S^{n-1}, i.e., A and B are disjoint and each one is equidecomposable to S^{n-1}. Define $A^* = \{(x_1, \dots, x_n, x_{n+1}) \in S^n \mid (x_1, \dots, x_n)/\sqrt{\sum_{i=1}^{n} x_i^2} \in A\}$ and B^* in a similar way. Then it is easy to see that $S^n \backslash \{(0, \dots, 0, \pm 1)\} = A^* \cup B^*$ is a paradoxical decomposition. In a similar way to the proof of (2.1.13) we deduce that $S^n \backslash \{(0, \dots, 0, \pm 1)\}$ is equidecomposable to S^n. Hence we have

Corollary 2.1.15. *S^n is $SO(n+1)$-paradoxical for every $n \geq 2$.*

In fact, for odd $n \geq 3$ the last corollary follows also from the following Theorem of Deligne and Sullivan and Example 2.1.6(ii). This Theorem gives more information on the existence of «minimal decompositions», but we shall not go into details here (see [DS], [Wa], [Bo2] and the references therein).

Theorem 2.1.16. ([DS]) *For odd $n \geq 3$, $SO(n+1)$ has a free subgroup of rank 2 which acts freely on S^n.*

For our discussion later on we will need a stronger form of the Hausdorff-Banach-Tarski paradox:

Theorem 2.1.17. *If $n \geq 2$, then any two subsets of S^n, each of which has non-empty interior, are $SO(n+1)$-equidecomposable. In particular, S^n is equidecomposable with every subset of it whose interior is not trivial.*

For proving this we need to extend slightly the notion of equidecomposable: we say for positive integers n and m and two subsets A and B of X that nA is equidecomposable with mB if one can decompose A in n ways and rebuild m copies of B (using translations by the action of G). We write $nA \sim mB$. So «X is paradoxical» just means that $X \sim 2X$. Moreover, the cancellation law works here, i.e., if $nA \sim nB$ then $A \sim B$. To see this look at the bi-partite graph whose set of inputs is A and outputs B, where every point of A is connected with the n points associated to it by the maps establishing the relation $nA \sim nB$.

More precisely: when decomposing A in n (different) ways and rebuilding n copies of B, we associate every $a \in A$ with $\varphi_1(a), \dots, \varphi_n(a) \in B$, where every $b \in B$ is obtained this way n times. So every $a \in A$ is connected by n edges of n different colors to points in B. It is possible, however, that a point $b \in B$ would receive two or more edges of the same color, but the total number of edges targeting at b is exactly n. Now, by the infinite version of König's Theorem (cf. [Or]), such a graph has a perfect matching. Namely, there is a one-to-one correspondence $\psi : A \to B$ such that for every $a \in A, \psi(a) = \varphi_i(a)$ for some $i \in \{1, \dots, n\}$. Let $A_i = \{a \in A \mid \psi(a) = \varphi_i(a)\}$. Then $A_i \sim \varphi_i(A_i) = \psi(A_i)$ and hence the disjoint unions $A = \bigcup_{i=1}^{n} A_i \sim \bigcup_{i=1}^{n} \psi(A_i) = B$ are also equivalent.

Now to prove Theorem 2.1.17, it suffices to prove its second part, i.e., that S^n is equidecomposable with any subset A with a non-trivial interior. Since sufficiently many copies of A cover $X = S^n$ we know that $X \precsim kA$, while $X \sim 2X \sim kX$ by Corollary 2.1.15. So, $kA \leq kX \sim X \precsim kA$ and hence $kA \sim kX$ by (2.1.2) (which can be easily seen to hold also with the slightly more general notion of equidecomposability). Thus $A \sim X$ and the theorem is proven. □

We conclude this section by mentioning that S^1 is not $SO(2)$-paradoxical. This will be shown in the next section.

2.2 Invariant Measures

An immediate corollary of the Hausdorff-Banach-Tarski paradox (2.1.17) is:

Proposition 2.2.1. *If $n \geq 2$, there is no finitely additive, rotation invariant measure of total measure 1 defined for all subsets of S^n.*

In general, the existence of a paradox eliminates the existence of an invariant measure defined for all subsets. It is interesting to mention Tarski's Theorem (cf. [Wa, Ch. 9]) which gives the converse:

Theorem 2.2.2. (Tarski) *Suppose G acts on X. Then there is a finitely additive, G-invariant measure $\mu : P(X) \to [0,1]$ with $\mu(X) = 1$ if and only if X is not G-paradoxical. ($P(X)$ denotes the set of all subsets of X.)*

Contrary to (2.2.1), for S^1 there is such an invariant measure. This is exactly the amenability of S^1 as a discrete group. We shall begin with a less standard definition of amenability, which is known to be equivalent to it by Følner Theorem.

Definition 2.2.3. *A locally compact group G is called amenable if it satisfies the following condition:*

(F) Given $\varepsilon > 0$ and a compact set $K \subset G$, there is a Borel set $U \subseteq G$ of positive finite (left Haar) measure $\lambda(U)$ such that $\frac{1}{\lambda(U)}\lambda(xU\Delta U) < \varepsilon$ for all $x \in K$. (Here $A\Delta B$ means $(A\backslash B) \cup (B\backslash A)$).

In particular, a discrete group is amenable if for every $\varepsilon > 0$ and every finite subset K there exists a non-empty finite set U such that $\frac{1}{|U|}|KU\Delta U| < \varepsilon$ (where $|A|$ stands for the number of elements in A).

To state this in a more geometric-combinatorial way: Recall that if G is a discrete group and K a subset, the Cayley graph $X(G;K)$ of G with respect to K is defined as a graph whose vertices are the elements of G and every such $g \in G$ is connected with $gk \in G$ for every $k \in K$. The condition of G being amenable is precisely that for every $\varepsilon > 0$ and every such K, the graph $X(G;K)$ has a finite subset U of vertices whose boundary ∂U satisfies $|\partial U| < \varepsilon|U|$. Moreover, if G is finitely generated, it is not difficult to see that once this happens to one *set of generators* K_0, it happens to every finite subset K.

The reader should also notice that the condition of amenability is precisely the opposite to the «expander» property discussed in Chapter 1. This is a first hint to the connection between the two topics.

It is easy to see that if every finitely generated subgroup of a discrete group G is amenable, then G is amenable. Finitely generated abelian groups are amenable, and therefore every abelian group is amenable; in particular, S^1 is amenable as a discrete group. (Indeed, it is amenable also as a topological group, just as every compact group is, since we can take U to be G). Moreover, extensions of amenable by amenable are amenable so solvable groups are also amenable.

Definition 2.2.4. *Let G be a locally compact group and $L^\infty(G)$ the (equivalence classes of) essentially bounded real-valued measurable functions (i.e., bounded outside a set of zero Haar measure). An invariant mean on G is a linear functional $m : L^\infty(G) \to \mathbb{R}$ satisfying:*

(i) $m(f) \geq 0$ *if* $f \geq 0$.

(ii) $m(\chi_G) = 1$ *where* χ_G *is the constant function* 1 *on* G.

(iii) $m(g\cdot f) = m(f)$ *for every* $g \in G$ *and* $f \in L^\infty(G)$ *where* $(g\cdot f)(x) = f(g^{-1}x)$.

Note that for a discrete group G, $L^\infty(G)$ is the space of *all* bounded functions on G. If such an m exists, then by defining, for a subset A of G, a measure $\mu(A) = m(\chi_A)$, we obtain a finitely additive G-invariant measure defined on all subsets of G.

Proposition 2.2.5. *If a locally compact group is amenable (i.e., satisfies condition (F) of 2.2.3), then it has an invariant mean.*

Remark. The two properties are really equivalent (see [Gl]). In fact, it is more common to define an amenable group as one having an invariant mean, but for our needs the one direction suffices.

Proof. Let H be the linear subspace of $X = L^\infty(G)$ spanned by the functions of the form $g \cdot f - f$ where $g \in G$ and $f \in X$. Clearly $\nu \in X^*$ is G-invariant if it annihilates H.

Lemma 2.2.6. *If $h \in H$ then $\|h\|_\infty \geq 0$ where $\|h\|_\infty = \operatorname{ess\,sup}_{x \in G} h(t) = \inf\{\alpha \mid h(t) \leq \alpha$ outside a null set$\}$.*

Proof. Let $h \in H$. h is of the form $h = \sum_{i=1}^{n}(k_i f_i - f_i)$ where $K = \{k_1, \ldots, k_n\}$ is some finite subset of G. Let U be as in (2.2.3) with respect to $K^{-1} = \{k_i^{-1}\}$ and $\varepsilon > 0$. Let $T(x) = \int_U h(ux)d\lambda(u)$. If $\|h\|_\infty = -\delta$ where $\delta > 0$, then $|T(x)| \geq \delta \cdot \lambda(U)$. On the other hand

$$\begin{aligned} T(x) &= \sum_{i=1}^{n} \int_U (f_i(k_i^{-1}ux) - f_i(ux))d\lambda(u) \\ &= \sum_{i=1}^{n} \Big(\int_{k_i^{-1}U} f_i(ux)d\lambda(u) - \int_U f_i(ux)d\lambda(u) \Big). \end{aligned}$$

Since $\lambda(k_i^{-1}U \Delta U) < \varepsilon\lambda(U)$, we have:

$$\delta\lambda(U) \leq |T(x)| \leq n \cdot \varepsilon \cdot \lambda(U) \cdot \max\{\| f_i \|_{L^\infty}\}.$$

As n and h are fixed and we can choose ε arbitrarily small, we come to a contradiction. $\square$

Remark 2.2.7. For the proof of (2.2.6) we could use $x = e$. The fact that x can be chosen arbitrarily can be used to prove the following stronger version, which is needed later.

If $h \in H$ and A is a G_δ-dense subset (or if A is a Lebesgue measurable subset of measure 1) of $G = S^1$, then $\|h\|_\infty \geq 0$.

To see this, note that in our case S^1 is amenable as a discrete group. We can therefore choose U in the proof of (2.2.6) to be a finite set. Then continue the proof word by word just choosing x such that Ux and $K^{-1}Ux$ are all in A, which is possible since the intersection of finitely many G_δ dense subsets is non-empty.

To complete the proof of (2.2.5): Take the subspace $Y = H \oplus \mathbb{R}\chi_G$ and define a linear functional $\nu : Y \to \mathbb{R}$ by $\nu(h + C\chi_G) = C$. By (2.2.6),

$$\| h + C\chi_G \|_\infty = \operatorname*{ess\,sup}_{t \in G} (h + C_{\chi_G}) \geq C = \nu(h + CX_G).$$

Thus, the Hahn-Banach Theorem (cf. [Ru2, Ch. 3]) ensures that we may extend ν to a linear functional m on X satisfying $|m(f)| \leq \| f \|_\infty$. This is the invariant mean we have looked for. □

Corollary 2.2.8. *There is a finitely additive, rotationally invariant measure of total measure 1 defined for all subsets of S^1. S^1 is, therefore, not paradoxical.*

Before turning to the Ruziewicz problem, we recall the following standard:

Proposition 2.2.9. *The Lebesgue measure λ is the unique rotation invariant countably additive measure defined on the set of all Lebesgue measurable subsets of S^n such that $\lambda(S^n) = 1$.*

Proof. This follows from the standard procedure of its construction: λ is determined uniquely on open balls in S^n since it is determined by covering properties of balls. By countably additivity it is determined on the σ-algebra generated by the closed and open subsets, i.e., the Borel sets. Then the Lebesgue measurable sets are obtained by completing the measure, i.e., by «adding» to the σ-algebra all subsets of zero measure sets. The monotonity of a measure implies that λ is still unique. □

Ruziewicz posed the question whether λ is still unique among the *finitely* additive invariant measures of S^n.

For the case $n = 1$, Banach [Ba] answered in negative, essentially proving along the way our (2.2.5) above. In fact, the construction in (2.2.5) is flexible enough to allow:

Proposition 2.2.10. *Let A be a G_δ-dense subset of S^1 such that $\lambda(A) = 0$. Then there exists a finitely additive invariant measure m defined on all subsets of S^1 such that $m(A) = 1$.*

Proof. We look again at the proof of (2.2.5) this time for the *discrete* group $G = S^1$. So ess sup is simply sup. Take now $Y = H \oplus \mathbb{R}\chi_G \oplus \mathbb{R}\chi_A$. This is indeed a direct sum, for if $\chi_A \in H \oplus \mathbb{R}\chi_G$ then $\chi_A = h + \alpha\chi_G$, so $h = \chi_A - \alpha\chi_G$ is constant on A. The same applies to $-h$ and hence $h = 0$ on A by (2.2.7), and it is equal $-\alpha$ on A^c, which is also a contradiction to (2.2.7). Define now $\nu(h+\alpha\chi_G+\beta\chi_A) = \alpha+\beta$. Then: $\| h+\alpha\chi_G+\beta\chi_A \| \geq \sup_{x\in A} |h+\alpha\chi_G+\beta\chi_A| \geq \alpha+\beta$ again by (2.2.7). As in the proof of (2.2.5), we can use Hahn-Banach Theorem to get m with $m(\chi_A) = 1$. □

The last Proposition shows the existence of a finitely additive measure which is totally different from λ – the Lebesgue measure. A measure μ is said to be absolutely continuous with respect to λ if $\mu(A) = 0$ whenever $\lambda(A) = 0$. If μ is such an invariant measure, then by the standard procedure to define an integral out of a measure μ gives rise to an invariant mean on $L^\infty(G)$, and vice versa: any invariant mean on $L^\infty(G)$ defines an absolutely continuous finitely additive invariant measure on the Lebesgue subsets of S^1. The next proposition establishes, therefore, that λ is not unique even among the absolutely continuous measures.

Proposition 2.2.11. *There exists an invariant mean I on $L^\infty(S^1)$ (S^1 with the usual topology) which is different than the Haar integral.*

Proof. Again the proof is similar to that of (2.2.5) and (2.2.10). This time take A to be an open dense subset of measure $\lambda(A) < 1$, and $B = S^1 \backslash A$. Extend the functional $\nu : H \oplus \mathbb{R}\chi_G \oplus \mathbb{R}\chi_B \to \mathbb{R}$, $\nu(h+\alpha\chi_G+\beta\chi_B) = \alpha$ to a functional I on $L^\infty(G)$ satisfying $I(\chi_B) = 0 \neq \int_B 1 d\lambda = 1 - m(A)$. □

Finally, we show, already at this point, that the situation in S^n for $n \geq 2$ is less flexible:

Proposition 2.2.12. *Let ν be a rotation invariant, finitely additive measure defined on Lebesgue subsets of S^n, $n \geq 2$. Then ν is absolutely continuous with respect to λ – the Lebesgue measure.*

Proof. Let E be a subset of S^n with $\lambda(E) = 0$. We have to show that $\nu(E) = 0$. Note first that if D_i is a sequence of open discs in S^n such that diameter$(D_i) \to 0$, then $\nu(D_i) \to 0$, because m_i, the number of disjoint translations of D_i in S^n, is going to infinity. Now by the Hausdorff-Banach-Tarski paradox (2.1.17), S^n is equidecomposable to D_i and E is therefore equidecomposable to a subset E_i of D_i. Now, *every subset of a zero measure set is measurable* and hence E_i is also measurable since the pieces of it, which are translations of pieces of E, are all Lebesgue measurable. Thus $\nu(E) = \nu(E_i) \leq \nu(D_i)$ and therefore $\nu(E) = 0$. □

Later on we will show, as promised, that for $n \geq 2$, $\nu = \lambda$, i.e., there are no «exotic» measures, but this will require much more work.

2.3 Notes

1. The Hausdorff-Banach-Tarski paradox is much more general than the case of S^n presented here. But it is based essentially on the same ingredients. A very detailed treatment is given in S. Wagon's book [Wa], where one can learn more about the history and the various contributors to the subject. As this subject is treated very intensively in the literature, we have not made any attempt to cover it here. We merely brought, in the shortest self-contained way we could (except of (2.1.7)), the case of S^n, which is the only case we need.
2. A similar remark applies to invariant measures: Banach [Ba] proved (2.2.8). It was Von-Neumann who defined amenable groups and realized their importance. Our treatment is closer to [Ru1]. We were also benefited from Sarnak [Sa2]. The standard reference on amenable groups is Greenleaf [Gl] and more recent ones are [P] and [Pie]. Wagon [Wa] also contains a lot of information, especially in connection to paradoxes.

3 Kazhdan Property (T) and its Applications

3.0 Introduction

Let G be a locally compact group. The set of equivalence classes of unitary (irreducible) representations is endowed with a topology, called the Fell topology. G is said to have property (T) if the trivial one-dimensional representation ρ_0 of G is isolated, in this topology, from all the other irreducible unitary representations of G. If Γ is a lattice in G, i.e., a discrete subgroup of finite co-volume, then Γ has property (T) if and only if G does. Simple Lie groups of rank ≥ 2, as well as their lattices, do have this property.

In this chapter we present this property and the results just mentioned. We then apply them to our problems: First, in §3.3 we show how one can use a discrete group Γ with property (T) to make explicit constructions of expanders. In fact, fixing a finite set S of generators of Γ, the Cayley graphs $X(\Gamma/N; S)$, when N runs over the finite index normal subgroups of Γ, form a family of expanders.

Secondly, it is shown that λ, the Lebesgue measure on S^n, is the unique finitely additive invariant measure on S^n(i.e., an affirmative answer to Ruziewicz problem) if there is a countable group Γ in $SO(n+1)$, such that all irreducible Γ-subrepresentations of $L_0^2(S^n) = \{f \in L^2(S^n) \mid \int f d\lambda = 0\}$ are bounded away from the trivial representation. This happens, in particular, if $SO(n+1)$ contains a countable dense subgroup Γ with property (T). Indeed, for $n \geq 4$, $SO(n+1)$ contains such a subgroup, and hence the Ruziewicz problem is answered affirmatively in this chapter for $n \geq 4$. For $n = 2, 3$ the answer is also affirmative, but for this we will have to wait until chapter 7. Here we prove that $SO(3)$ and $SO(4)$ do not contain a dense countable group with property (T). Hence a different tool is needed.

3.1 Kazhdan property (T) for semi-simple groups

We begin this section by introducing the Fell topology ([Fe]) on the unitary dual.

Definition 3.1.1. *Let G be a locally compact (separable) group and $\rho : G \to U(H)$ a continuous unitary representation of G on a (separable) Hilbert space. $\tilde{G}$ (resp: $\hat{G}$) denotes the set of equivalent classes of the unitary (resp: irreducible unitary) representations of G.*
(a) For every vector $v \in H$ with $\| v \| = 1$ we associate a coefficient of ρ. This is the function on $G : g \mapsto \langle v, \rho(g)v \rangle$ where $\langle\ ,\ \rangle$ is the scalar product in the Hilbert space H.
(b) For two representations σ and ρ, we say that ρ is weakly contained in σ, denoted $\rho \propto \sigma$, if every coefficient of ρ is a limit, uniformly on compact sets of G, of coefficients of σ.

(c) We define open neighborhoods of a representation (H, ρ) in $\hat{G}$ as follows: Let K be a compact subset of G, $\varepsilon > 0$ and $v \in H$ of norm one. Let:

$$W(K, \varepsilon, v) = \left\{ \begin{array}{l|l} (H', \sigma) \in \hat{G} & \exists v' \in H' \text{ of norm one such that} \\ & |\langle v, \rho(g)v \rangle - \langle v', \sigma(g)v' \rangle| < \varepsilon, \ \forall g \in K \end{array} \right\}.$$

This determines a topology on $\hat{G}$ (and $\tilde{G}$) (see [HV], [Zi1], [Fe], [M7] for details).

We spell out the above definition for the special case $\rho = \rho_0$, the one-dimensional trivial representation:

Proposition 3.1.2. *$\rho_0 \propto \sigma$ if and only if for every $\varepsilon > 0$ and every compact subset K of G, there exists $v \in H(\sigma)$ such that $\| v \| = 1$ and $|\langle v, \sigma(g)v \rangle - 1| < \varepsilon$, or equivalently $\| \sigma(g)v - v \| < \sqrt{2\varepsilon}$ for every $g \in K$. In this situation we say, for obvious reasons, that σ has «almost invariant vectors».*

Definition 3.1.3. *G is called a Kazhdan group or is said to have property T if ρ_0 is an isolated point in $\hat{G}$ with the Fell topology. This means that some neighborhood of ρ_0 contains only ρ_0. In other words: G is a Kazhdan group if there exist an $\varepsilon > 0$ and a compact subset K of G such that for every nontrivial irreducible representation (H, ρ) of G and every vector $v \in H$ of norm one, $\| \rho(k)v - v \| > \varepsilon$ for some k in K.*

An equivalent formulation is as follows (see [HV], [Zi1], [Ka] or [Wan1, 1.5] for details about this point as well as any other unexplained points in this section):

G has property T if and only if every unitary representation which has almost invariant vectors contains also a nonzero invariant vector, or equivalently if any unitary representation of G which weakly contains the trivial representation contains a nonzero invariant vector.

Example 3.1.4. (A) If G is a compact group, then G is Kazhdan. Indeed, let (H, ρ) be a representation of G containing almost invariant vectors. Let v be a vector in H of norm one such that $\| \rho(g)v - v \| < \frac{1}{2}$. Then $\bar{v} = \int_G \rho(g)v dg$ is a G-invariant vector and it is nonzero since $\| \bar{v} - v \| < \frac{1}{2}$.

(B) Let $G = \mathbb{R}$ (or $\mathbb{Z}$), then the left regular representation $L : G \to U(L^2(G))$ defined by $L(g)f(x) = f(g^{-1}x)$ weakly contains ρ_0 but does not contain ρ_0. Indeed, for every compact (and therefore bounded) subset K of $\mathbb{R}$ and every $\varepsilon > 0$ we can take a sufficiently large interval I of length ℓ, say, such that $\lambda((K+I)\Delta I) < \varepsilon\lambda(I)$ where λ is the Lebesgue measure. This shows that $1/\sqrt{\ell}\chi_I$ is ε-invariant under K. On the other hand, there is no invariant function since the constant function is not in $L^2(G)$. (It will be in $L^2(G)$ if and only if G is a compact group). It is clear that the same argument works for any amenable group. We thus have proved the «only if» part of the following:

Theorem 3.1.5. (Hulanicki [Hul]) *A group G is amenable if and only if the (left) regular representation of G weakly contains the trivial representation ρ_0.*

For future use, notice that the «if» part amounts to prove that if there are «almost invariant functions» then there are also almost «invariant subsets» (see also the proofs of (3.3.7) and (4.2.4)).

Proof of 3.1.5. Let G be a locally compact group with Haar measure λ. We say that G satisfies Property (F) if for every compact subset $K \subseteq G$ and every $\varepsilon > 0$ there exists a Borel subset $U \subseteq G$ with $0 < \lambda(U) < \infty$ such that $\lambda(kU\Delta U) < \varepsilon\lambda(U)$ for every $k \in K$. G is said to have property (F^*) if for every compact K, $\varepsilon > 0$ and $\delta > 0$ there exists a Borel U and a subset $N \subseteq K$ with $\lambda(N) < \delta$ such that $\lambda(kU\Delta U) < \varepsilon\lambda(U)$ for every $k \in K \setminus N$.

Theorem 3.1.5 is contained in the following:

Claim. *Let G be a locally compact group. Then the following three conditions are equivalent.*

(i) G has property (F), i.e., G is amenable.

(ii) G has property (F^*).

(iii) The trivial representation ρ_0 is weakly contained in the left regular representation L_G of G on $L^2(G)$.

Proof. (i) implies (iii) is the easy part of Theorem 3.1.5, which was proved in Example 3.1.4(B). (i) implies (ii) is clear. We will prove now that (iii) implies (ii) and then that (ii) implies (i).

We have to prove that if $\rho_0 \propto L_G$ then G has property (F^*). Therefore, let K be a compact subset of G, $\varepsilon > 0$ and $\delta > 0$. Let $\varepsilon_0 = \frac{\varepsilon^2\delta^2}{4\lambda^2(K)}$ and f a function in $L^2(G)$ with $\int_G |f|^2 \, \mathrm{d}\lambda = 1$ and $\int_G |kf - f|^2 \, \mathrm{d}g < \varepsilon_0$ for every $k \in K$. Let $F = f^2$ then $F \in L^1(G)$ and

$$\|kF - F\|_{L^1} = \int_G |kf^2 - f^2| \, \mathrm{d}\lambda = \int_G |(kf)^2 - f^2| \, \mathrm{d}\lambda =$$

$$\int_G |(kf - f)(kf + f)| \, \mathrm{d}\lambda \leq \left(\int_G |kf - f|^2 \, \mathrm{d}\lambda\right)^{\frac{1}{2}} \cdot \left(\int_G |kf + f|^2 \, \mathrm{d}\lambda\right)^{\frac{1}{2}} <$$

$$(\varepsilon_0)^{\frac{1}{2}} \|kf + f\|_{L^2} \leq (\varepsilon_0)^{\frac{1}{2}} (\|kf\|_{L^2} + \|f\|_{L^2}) = 2(\varepsilon_0)^{\frac{1}{2}}.$$

We thus have a function F in $L^1(G)$ with $\int F \, \mathrm{d}\lambda = 1$ which is almost invariant.

For $0 < a \in \mathbb{R}$ define the subset U_a of G to be $U_a = \{x \in G \mid |F(x)| \geq a\}$. Then $\int_G F \, \mathrm{d}\lambda = \int_0^\infty \lambda(U_a) \, \mathrm{d}a$.

(This follows from Fubini Theorem but is best illustrated by Figure 1).

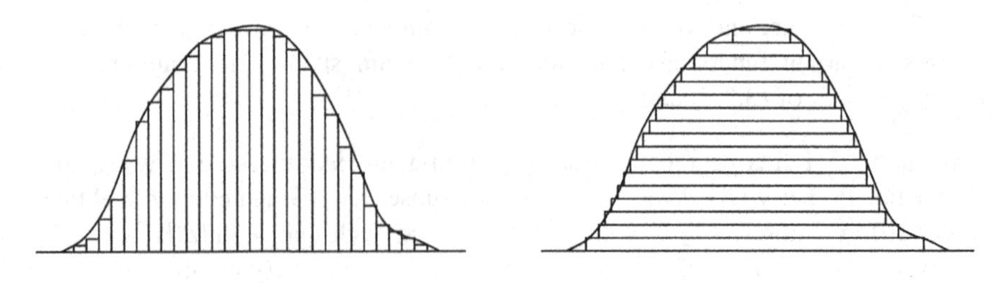

Fig. 1

Similarly $\|kF - F\|_{L^1} = \int_G |kF - F|\, d\lambda = \int_0^\infty \lambda(kU_a \Delta U_a)\, da$ (see Figure 2).

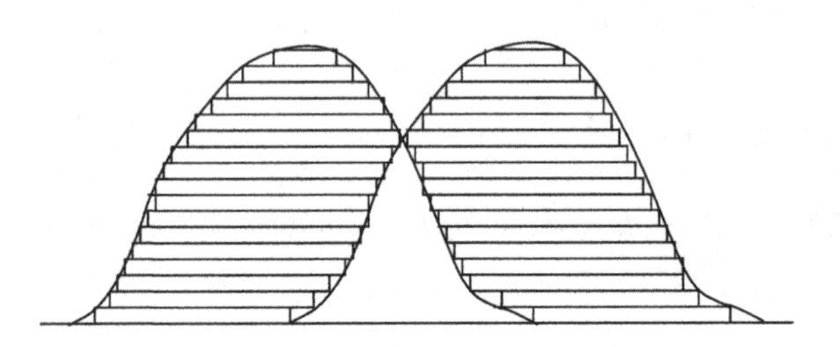

Fig. 2

Thus:

$$\int_0^\infty \lambda(kU_\alpha \Delta U_\alpha)\, da < 2\varepsilon_0^{\frac{1}{2}} = 2\varepsilon_0^{\frac{1}{2}} \cdot \int_0^\infty \lambda(U_a)\, da.$$

By integration on K,

$$\int_K \int_0^\infty \lambda(kU_a \Delta U_a)\, da\, dk < 2(\varepsilon_0)^{\frac{1}{2}} \lambda(K) = \varepsilon\delta,$$

i.e., $\int_0^\infty \int_K \lambda(kU_a \Delta U_a)\, dk \quad da \le \varepsilon \cdot \delta \int_0^\infty \lambda(U_a)\, da$. Thus, there exists $a \in \mathbb{R}$ such that

$$\int_K \lambda(kU_a \Delta U_a)\, dk \le \varepsilon \cdot \delta \lambda(U_a).$$

It follows that the measure of those $k \in K$ for which $\lambda(kU_a \Delta U_a) > \varepsilon \lambda(U_a)$ is less than or equal to δ. This proves that G has property (F^*).

It remains to prove that (F^*) implies (F). We remark first that for discrete groups they are clearly equivalent since we can take $\delta < \frac{1}{|K|}$.

To prove $(F^*) \Rightarrow (F)$: Let $K \subseteq G$ be compact with $\lambda(K) > 0$. Let $A = K \cup K \cdot K$. Then for every $k \in K$, $\lambda(kA \cap A) \ge \lambda(kK) = \lambda(K)$. Let $\delta = \frac{1}{2}\lambda(K)$.

Then for every $N \subset A$ such that $\lambda(A \setminus N) < \delta$ and for every $k \in K$

$$2\delta \leq \lambda(K) \leq \lambda(kA\cap A) \leq \lambda(kN\cap N)+\lambda(A\setminus N)+\lambda(k(A\setminus N)) < \lambda(kN\cap N)+2\delta.$$

Hence, $\lambda(kN \cap N) > 0$, and so $k \in NN^{-1}$ and therefore $K \subseteq NN^{-1}$.

Now, use property (F^*) for $\frac{\varepsilon}{2}, \delta$ and A. So, there is U with $0 < \lambda(U) < \infty$ and $N \subset A$ such that $\lambda(A \setminus N) < \delta$ and $\lambda(kU\Delta U) < \frac{\varepsilon}{2}\lambda(U)$ for every $n \in N$. So, for $n_1, n_2 \in N$ we have:

$$\begin{aligned}\lambda(n_1n_2^{-1}U\Delta U) &\leq \lambda(n_2^{-1}U\Delta U) + \lambda(U\Delta n_1^{-1}U)\\ &= \lambda(n_2U\Delta U) + \lambda(n_1U\Delta U) < \varepsilon\lambda(U)\end{aligned}$$

and thus $\lambda(kU\Delta U) < \varepsilon\lambda(U)$ for every $k \in K \subseteq NN^{-1}$. □

From the above one can also deduce:

Corollary 3.1.6. *If G is amenable and has property T, then G is compact.*

Proof. The regular representation contains the trivial representation if and only if G is compact. □

A quotient of a Kazhdan group is also Kazhdan. One therefore deduces that the commutator quotient $G/[G,G]$ of a Kazhdan group G is compact and $G/[G,G]$ is finite if G is discrete. Thus

Proposition 3.1.7. *A non-abelian free group F is neither Kazhdan nor amenable.*

Proof. It is not Kazhdan by the above. It is not amenable either by checking directly or by recalling from (2.1.6) that F is paradoxical and therefore does not have an invariant mean. □

Kazhdan introduced in [Ka] property (T) as a tool for studying discrete subgroups of Lie groups of finite co-volume. Recall that we say that a discrete subgroup of G is a *lattice* if G/Γ carries a G-invariant finite measure. The following proposition summarizes the connection between the Fell topologies of the unitary dual of Γ and of G.

Proposition 3.1.8. *(a) Let G be a locally compact group and H a closed subgroup. Then the restriction map $\tilde{G} \xrightarrow{\text{res}} \tilde{H}$ given by $\rho \mapsto \rho\,|_H$ is a continuous map. Moreover, the induction map $\pi \mapsto \operatorname{Ind}_H^G(\pi)$ is a continuous map from $\tilde{H}$ to $\tilde{G}$.*

(b) If Γ is a lattice in G, then: (i) G is amenable if and only if Γ is amenable, and (ii) G has property (T) if and only if Γ has.

We omit the proof which can be found in [Fe] and the above-mentioned references. Notice that part (a) says that if $\pi' \propto \pi$ where π' and π are representations of H, then $\operatorname*{Ind}_H^G(\pi') \propto \operatorname*{Ind}_H^G(\pi)$. This with the aid of (3.1.5) suffices to prove that if Γ is amenable then G is amenable.

An immediate corollary is:

Proposition 3.1.9. $G = SL_2(\mathbb{R})$ *is neither amenable nor has property* (T).

Proof. As it is well known (see also (3.2.1)), $SL_2(\mathbb{Z})$ is a lattice in G and it has a free subgroup F of finite index (for example, the group generated by $\begin{pmatrix} 1 & 2 \\ 0 & 1 \end{pmatrix}$ and $\begin{pmatrix} 1 & 0 \\ 2 & 1 \end{pmatrix}$). Thus F is also a lattice in G. So (3.1.7) and (3.1.8) prove (3.1.9). $\square$

It is more difficult to prove that some groups do have property (T). Before doing this we shall extend slightly the definition:

Definition 3.1.10. *If R is a subset of $\hat{G}$, we say that G has property $(T : R)$ if ρ_0 is isolated in $R \cup \{\rho_0\}$.*

Proposition 3.1.11. *Let $H = \mathbb{R}^2 \rtimes SL_2(\mathbb{R})$, i.e., the semi-direct product with the standard action of $SL_2(\mathbb{R})$ on $\mathbb{R}^2$. Let $R = \{\rho \in \hat{H} | \rho|_{\mathbb{R}^2}$ is nontrivial$\}$. Then H has property $(T : R)$.*

Proof. Let ρ be an irreducible unitary representation of H. The restriction of ρ to $\mathbb{R}^2$ is a unitary representation of the abelian group $\mathbb{R}^2$ and thus a direct sum (or integral) of one-dimensional characters of $\mathbb{R}^2$. Let χ be a character appearing there and let M be its stabilizer in $SL_2(\mathbb{R})$ that acts also on $\hat{\mathbb{R}}^2$ the group of characters. By Mackey's Theorem (cf. [Zi1, 7.3.11]) $\rho = \operatorname*{Ind}_{\mathbb{R}^2 \rtimes M}^{H}(\chi\sigma)$ where σ is some irreducible representation of M and $\chi\sigma$ is the representation of $\mathbb{R}^2 \rtimes M$ defined by $(\chi\sigma)(r,m) = \chi(r)\sigma(m)$. Now, if in addition ρ is in R, then χ is nontrivial and thus M is a proper subgroup of SL_2 (in fact M is conjugate to $N = \left\{\begin{pmatrix} 1 & t \\ 0 & 1 \end{pmatrix}\right\}$ since all nontrivial characters of $\mathbb{R}^2$ are conjugate under the $SL_2(\mathbb{R})$ action). $M_1 = \mathbb{R}^2 \rtimes M$ is nilpotent and, therefore, amenable. By 3.1.5 this means that $\rho_0 \propto L_{M_1}$, where L_{M_1} is the left regular representation of M_1, and thus

$$\chi\sigma = (\chi\sigma \otimes \rho_0) \propto (\chi\sigma) \otimes L_{M_1} \simeq \dim(\chi\sigma) \cdot L_{M_1}.$$

(The last isomorphism follows from the identification, for any representation (W,π), $L^2(G)\otimes W \xrightarrow[\sim]{\theta} L^2(G,W)$ given by:

$$\theta(f\otimes v)(g)=f(g)(\pi(g)v)$$

where $$L^2(G,W)=\{f:G\to W \mid \int_G \| f(g) \| \, dg<\infty\}.)$$

Thus (3.1.8) (and the remark afterward) implies that

$$\rho=\operatorname*{Ind}_{M_1}^{H}(\chi\sigma)\propto\infty\cdot\operatorname*{Ind}_{M_1}^{H}(L_{M_1})=\infty\cdot L_H.$$

Now, if $\rho_0\propto\rho$ then $\rho_0\propto L_H$, which is a contradiction to (3.1.5) and (3.1.9). Thus ρ does not weakly contain the trivial representation, which implies that ρ_0 is bounded away from R in the Fell topology. □

Theorem 3.1.12. *$SL_3(\mathbb{R})$ has property (T).*

Lemma 3.1.13. *Let $E=SL_2(\mathbb{R})$ and $N=\left\{\begin{pmatrix}1&t\\0&1\end{pmatrix} \mid t\in\mathbb{R}\right\}$. If ρ is a unitary representation of E, then every vector fixed by $\rho(N)$ is also fixed by $\rho(E)$.*

Proof. Suppose $\rho(n)v=v$ for every $n\in N$. Look at the function on E, $f(g)=\langle\rho(g)v,v\rangle$. We claim that this is a constant function on E. This will prove the lemma, since then $f(g)=f(e)=\langle v,v\rangle$ and so $\rho(g)v=v$. To see that f is constant, we argue in several steps:

Step 1. $f(n_1gn_2)=f(g)$ for every $n_1,n_2\in N$. Indeed, $f(n_1gn_2)=\langle\rho(n_1gn_2)v,v\rangle=\langle\rho(n_1)\rho(g)\rho(n_2)v,v\rangle=\langle\rho(n_1)\rho(g)v,v\rangle=\langle\rho(g)v,\rho(n_1)^{-1}v\rangle=\langle\rho(g)v,v\rangle=f(g)$.

Step 2. Identify E/N with $\mathbb{R}^2\backslash\{0\}$. This can be done since the action of E on $\mathbb{R}^2\backslash\{0\}$ is transitive with N being the stabilizer of $\binom{1}{0}$. So f, which is constant on the double cosets $N\backslash E/N$, can be thought of as a function on $\mathbb{R}^2\backslash\{0\}$ constant on N orbits. N acts like $\begin{pmatrix}1&t\\0&1\end{pmatrix}\binom{x}{y}=\binom{x+ty}{y}$. So the orbits of N in $\mathbb{R}^2\backslash\{0\}$ are the lines parallel to the x-axis and each point on the x-axis. Since f is constant on all the lines parallel to the x-axis and it is continuous, it is also constant on the x-axis. Restating this by means of the group elements, we have that if $p\in P=\left\{\begin{pmatrix}a&b\\0&d\end{pmatrix}\in E\right\}$ then $f(p)=f(e)$, i.e., the vector v is also invariant under P.

Step 3. If $p_1,p_2\in P$ then $f(p_1gp_2)=f(g)$ for every $g\in E$. This is proved exactly like Step 1, using the fact, established in Step 2, that P fixes v.

Step 4. Now identify E/P with the projective line $\mathbf{P}^1(\mathbb{R}) = \mathbb{R} \cup \{\infty\}$ via the action of E by $\begin{pmatrix} a & b \\ c & d \end{pmatrix}(z) = \frac{az+b}{cz+d}$. So f is defined on $\mathbf{P}^1(\mathbb{R})$ and constant on the P orbits. There are only two orbits: $\mathbb{R}$ and the point ∞. Again, by continuity, f is constant on the entire projective line and therefore constant on E.

This proves that v is fixed by E and the Lemma is proved. □

Lemma 3.1.14. *Let* $J = \left\{ \begin{pmatrix} 1 & 0 & s \\ 0 & 1 & t \\ 0 & 0 & 1 \end{pmatrix} \middle| \, s,t \in \mathbb{R} \right\} \subseteq G = SL_3(\mathbb{R})$. *For any unitary representation of G, each vector which is left fixed under J is left fixed under G.*

Proof. Let
$$E_1 = \left\{ A = \begin{pmatrix} a & 0 & b \\ 0 & 1 & 0 \\ c & 0 & d \end{pmatrix} \middle| \, a,b,c,d \in \mathbb{R} \text{ and } \det(A) = 1 \right\}$$
and
$$E_2 = \left\{ A = \begin{pmatrix} 1 & 0 & 0 \\ 0 & a & b \\ 0 & c & d \end{pmatrix} \middle| \, a,b,c,d \in \mathbb{R} \text{ and } \det(A) = 1 \right\}.$$

It is not difficult to see that E_1 and E_2 generate a dense subgroup of G. Let $N_i = E_i \cap J$ for $i = 1,2$. From Lemma 3.1.13 we deduce that a fixed vector of N_i is a fixed vector of E_i. We conclude that if v is fixed by J, it is fixed by both E_1 and E_2 and hence by G. □

Proof of 3.1.12. Let ρ be a unitary representation of $G = SL_3(\mathbb{R})$ and assume ρ weakly contains ρ_0. Let $H = \left\{ \begin{pmatrix} a & b & r \\ c & d & s \\ 0 & 0 & 1 \end{pmatrix} \in G \right\} \simeq J \rtimes SL_2(\mathbb{R}) \simeq \mathbb{R}^2 \rtimes SL_2(\mathbb{R})$. By definition, ρ restricted to H also contains weakly the trivial representation. From (3.1.11) it follows that $\rho|_H$ (strongly) contains representations which are not in $R = \{\rho \in \hat{H} | \rho|_{\mathbb{R}^2} \text{ is nontrivial}\}$. Namely, it contains a vector invariant under $\mathbb{R}^2 \simeq J$. By (3.1.14) this vector is also invariant under G. This proves that G is a Kazhdan group. □

Going over the proof one sees that there is nothing special about $\mathbb{R}$, and the proof works equally well for every locally compact *non-discrete* field (the non-discreteness is needed to apply the continuity arguments in Lemma 3.1.13, and indeed $SL_3(\mathbb{Q})$ does not have property (T) since it is not a finitely generated group, while every countable Kazhdan group is finitely generated – see [Ka], [Wan1], [Zi1] or [HV]). Moreover the theorem is not limited to SL_3 but is true in a much greater generality. Recall that if $\underset{\sim}{G}$ is an algebraic group defined over a field

K then the K-rank of $\underset{\sim}{G}$ is the dimension of the maximal K-split tori in G, e.g., K-rank $(SL_n) = n-1$ for every field K, and $\mathbb{R}$-rank $(SO(n,1)) = 1$. Over different fields $\underset{\sim}{G}$ it might have different ranks, for example $\mathbb{C}$-rank $(SO(n,1)) = \left[\frac{n+1}{2}\right]$, or $\mathbb{R}$-rank $(SO(n)) = 0$, $\mathbb{C}$-rank $(SO(n)) = \left[\frac{n}{2}\right]$ and $\mathbb{Q}_p$-rank $(SO(n))$ depends on p but of course always at most equal to the $\mathbb{C}$-rank.

Theorem 3.1.15. *Let F be a local field and G the group of F-points of a simple algebraic group $\underset{\sim}{G}$ defined over F, of F-rank ≥ 2. Then G is a Kazhdan group.*

The proof requires some background in algebraic groups. The reader is referred to the above-mentioned references. He or she is encouraged to work out the case $G = SL_n(\mathbb{R}), n \geq 3$. In later sections we make use only of the cases $G = SL_2(\mathbb{R}) \ltimes \mathbb{R}^2$ or $SL_3(\mathbb{R})$ which we have proved in detail, or $G = SO(n,2)$ $(n \geq 3)$ and/or $G = SO_n(\mathbb{Q}_5) (= SO(n,0)(\mathbb{Q}_5) = SO(n,\mathbb{Q}_5))$ $(n \geq 5)$ which we have not proved. But a similar proof can be given for them using detailed information on their structure.

3.2 Lattices and arithmetic subgroups

In this section we describe a method to construct lattices in semi-simple Lie groups. This will be done by arithmetic means. By a remarkable theorem of Margulis (cf. [M7], [Zi1]), this is essentially the only method for groups of rank ≥ 2. The method of construction for some groups (e.g., $G = SO(n,2)$) is as important for us, for the application in §3.4, as the fact that such lattices exist. The reason is that the construction is made in a natural way, as lattices in $G \times SO(n+2)$, and then one takes projection into G. The fact that we can also take projection into the other factor $SO(n+2)$ is the key point for solving the Ruziewicz problem for S^{n+1}.

Later on we will also need lattices in p-adic Lie groups. So we now give a very general statement, but we will actually use only very specific examples.

Let S' be a finite set of primes in $\mathbb{Z}$ and $S = S' \cup \{\infty\}$. For every $p \in S$, we denote by $\mathbb{Q}_p$ the field of p-adic numbers where $\mathbb{Q}_\infty = \mathbb{R}$. Let $\mathbb{Z}_S$ be the ring of S-integers, i.e., the subring of $\mathbb{Q}$ consisting of all rational numbers whose denominators have all prime factors lying in S. $\mathbb{Z}_S$ is diagonally embedded in $\prod_{p \in S} \mathbb{Q}_p$ and is a discrete co-compact subring. Now let $\underset{\sim}{G}$ be a connected semi-simple algebraic $\mathbb{Q}$-group.

Theorem 3.2.1. $\Gamma = \underset{\sim}{G}(\mathbb{Z}_S)$ *is a lattice in* $G = \prod_{p \in S} \underset{\sim}{G}(\mathbb{Q}_p)$.

This theorem is due to Borel [Bo1], generalizing a previous result of him with Harish-Chandra [BHC] for the case $S = \{\infty\}$.

Example 3.2.2. (A) $\Gamma = SL_n(\mathbb{Z})$ is a lattice in $G = SL_n(\mathbb{R})$. This is the case $\underset{\sim}{G} = SL_n$ and $S = \{\infty\}$ in the theorem. But it is a classical result. The case of $\Gamma = SL_2(\mathbb{Z})$ can be proved directly by considering the symmetric space $\mathbf{H} = \{x + iy \mid x, y \in \mathbb{R}, y > 0\}$ of $SL_2(\mathbb{R})$ that is naturally identified with $SL_2(\mathbb{R})/SO(2)$ via the action of $SL_2(\mathbb{R})$ on $\mathbf{H}$ by Möbius transformations. This action is transitive, and $SO(2)$ is the stabilizer of the point i. A fundamental domain for Γ is the set $F = \{z \in \mathbb{H} \mid |Re\ z| \leq \frac{1}{2} \text{ and } |z| \geq 1\}$. F has a finite area with respect to the hyperbolic measure $\frac{dxdy}{y^2}$, which is the G-invariant measure of $\mathbf{H}$ induced by a Haar measure of G.

(B) Let $\underset{\sim}{G} = \underset{\sim}{S}O(n) = \{A \mid A$ is an $n \times n$ matrix with $A^tA = I$ and $\det A = 1\}$ and take $S = \{5, \infty\}$. Then $\mathbb{Z}_S = \mathbb{Z}[\frac{1}{5}]$ and $\Gamma = G(\mathbb{Z}[\frac{1}{5}])$ is a lattice in $\underset{\sim}{G}(\mathbb{Q}_5) \times SO(n)$.

(C) Let f be the quadratic form

$$f(x_0, \ldots, x_n) = x_0^2 + x_1^2 + \ldots + x_{n-2}^2 - \sqrt{2}x_{n-1}^2 - \sqrt{2}x_n^2$$

and $\underset{\sim}{G} = \underset{\sim}{G}(f)$ the group of $(n+1) \times (n+1)$ matrices preserving this form. Let $k = \mathbb{Q}(\sqrt{2})$ and $\mathcal{O} = \mathbb{Z}[\sqrt{2}]$, the ring of algebraic integers in k. Let σ be the non-trivial element of the Galois group $\mathrm{Gal}(k/\mathbb{Q})$; ${}^\sigma f$ is the form ${}^\sigma f = x_0^2 + x_1^2 + \ldots + x_{n-2}^2 + \sqrt{2}x_{n-2}^2 + \sqrt{2}x_n^2$ and ${}^\sigma \underset{\sim}{G}$ is the algebraic group preserving ${}^\sigma f$. Then $\Gamma = \underset{\sim}{G}(\mathcal{O})$ is a lattice in $\underset{\sim}{G}(\mathbb{R}) \times {}^\sigma \underset{\sim}{G}(\mathbb{R}) \simeq SO(n-1, 2) \times SO(n+1)$. This is also a special case of Theorem 3.2.1, but some explanation is in order. The group $\underset{\sim}{G}$, to begin with, is defined over k, but $\underset{\sim}{G} \times^\sigma \underset{\sim}{G}$ is defined over $\mathbb{Q}$. This is just the procedure called «restriction of scalars» described, for example, in [Zi1, 6.1.3]. It is also explained there that $\Gamma = \underset{\sim}{G}(\mathcal{O})$ corresponds to $(\underset{\sim}{G} \times^\sigma \underset{\sim}{G})(\mathbb{Z})$, and so it is a lattice. Also, we know that over $\mathbb{R}, f$ is equivalent to the quadratic form $\sum_{i=0}^{n-2} x_i^2 - x_{n-1}^2 - x_n^2$ while ${}^\sigma f$ is positive definite and equivalent to $\sum_{i=0}^{n} x_i^2$. $SO(n-1, 2)$ denotes the group of real matrices preserving the first one, and $SO(n+1)$ of those preserving the second. For more information and examples of arithmetic groups defined using number fields, the reader is referred to Mostow [Mo].

(D) Let $\underset{\sim}{D} = \underset{\sim}{D}(\alpha, \beta)$ be a quaternion algebra defined over $\mathbb{Q}$, i.e., for a ring R, $\underset{\sim}{D}(R) = \{x = a_0 + a_1 i + a_2 j + a_3 k \mid a_i \in R\}$ where α and β are rational numbers and $i^2 = -\alpha, j^2 = -\beta, k^2 = -\alpha\beta, ij = -ji = k$. The norm of x is $N(x) = a_0^2 + \alpha a_1^2 + \beta \alpha_2^2 + \alpha\beta a_3^2$. Let $\underset{\sim}{G} = \underset{\sim}{D}^*/Z$ be the algebraic group of the invertible elements of $\underset{\sim}{D}$ modulo its center. Then $\Gamma = \underset{\sim}{G}(\mathbb{Z}[\frac{1}{p}])$ is a lattice in

$\underset{\sim}{G}(\mathbb{R})\times\underset{\sim}{G}(\mathbb{Q}_p)$. As it is well known (cf. [Re]), $\underset{\sim}{D}$ splits for almost every prime, i.e., $\underset{\sim}{D}(\mathbb{Q}_p)\simeq M_2(\mathbb{Q}_p)$, and so $\underset{\sim}{G}=PGL_2(\mathbb{Q}_p)$. For example, if $\alpha=\beta=1$ then $\underset{\sim}{D}$ is the standard Hamiltonian quaternion algebra. This one splits for every $p\neq 2,\infty$, and $\Gamma=\underset{\sim}{G}(\mathbb{Z}[\frac{1}{p}])$ is in fact the group presented in the proof of Proposition 2.1.7. More precisely, Γ is embedded diagonally in $PGL_2(\mathbb{Q}_p)\times SO(3)$ (since $\underset{\sim}{G}(\mathbb{R})\simeq SO(3)$), and the free group in (2.1.7) is the congruence subgroup mod 2 in Γ projected to $SO(3)$. Γ is a lattice in $PGL_2(\mathbb{Q}_p)\times SO(3)$ by (3.2.1), and since $SO(3)$ is compact, its projection to $PGL_2(\mathbb{Q}_p)$ is a lattice there. We will come back to this lattice in Chapter 7.

Looking again into property (T), it is not difficult to prove that a direct product of finitely many Kazhdan groups is also Kazhdan. Each one of the groups $\underset{\sim}{G}(\mathbb{Q}_p)$ in Theorem 3.2.1 has property (T) if it is either compact or has $\mathbb{Q}_p$-rank ≥ 2. We can therefore deduce from (3.2.1), (3.1.8b) and (3.1.15):

Proposition 3.2.3. *If $\Gamma=\underset{\sim}{G}(\mathbb{Z}_S)$ as before and for every $p\in S$ (including $p=\infty$) $\mathbb{Q}_p$-rank$(\underset{\sim}{G}(\mathbb{Q}_p))\neq 1$, then Γ has property (T).*

Proof. The only thing we should add is that $\mathbb{Q}_p$-rank$(\underset{\sim}{G}(\mathbb{Q}_p))=0$ if and only if $\underset{\sim}{G}(\mathbb{Q}_p)$ is compact. This is a standard result for $\mathbb{R}$ and due to Bruhat and Tits in the other cases (cf. [Pr1]). □

Example 3.2.4. We will go over the examples presented in (3.2.2) and check whether they are Kazhdan:

(A) $SL_n(\mathbb{Z})$ is Kazhdan if and only if $n\geq 3$ (or $n=1$). This follows from (3.1.15) and (3.1.9).

(B) The $\mathbb{Q}_5$-rank of $\underset{\sim}{S}O(n)$ is $[\frac{n}{2}]$ (since $x^2\equiv -1$ has a solution in $\mathbb{Q}_5$). While the $\mathbb{R}$-rank of $SO(n)$ is 0, so $SO(n,\mathbb{Z}[\frac{1}{5}])$ has property (T) if $n\geq 5$. (Warning: If $n=4$, $SO(4,\mathbb{Q}_5)$ has rank 2, but it is *not* a simple algebraic group; in fact, it is a product of two rank one groups and does not have property (T).)

(C) The group $SO(p,q)$ preserving the form $\sum\limits_{i=1}^{p}x_i^2-\sum\limits_{i=p+1}^{p+q}x_i^2$ has $\mathbb{R}$-rank $\min(p,q)$, so $SO(n+1)=SO(n+1,0)$ has rank zero while $SO(n-1,2)$ has rank 2 (unless $n\leq 3$ where for $n=3$, $SO(2,2)$ is locally isomorphic to $PSL_2(\mathbb{R})\times PSL_2(\mathbb{R})$; for $n=2$, $SO(1,2)\simeq PSL_2(\mathbb{R})$; and for $n=1$, $SO(n-1,2)\simeq SO(2)$ is compact). So $\Gamma=\underset{\sim}{G}(\mathbb{O})$ has property (T) if and only if $n\neq 2,3$.

(D) $\underset{\sim}{G}(\mathbb{Q}_p)$ is either compact or isomorphic to $PGL_2(\mathbb{Q}_p)$. Thus $\Gamma=\underset{\sim}{G}(\mathbb{Z}[\frac{1}{p}])$ has property (T) if and only if $\underset{\sim}{D}$ is ramified at both ∞ and p, in which case Γ is a finite group.

Before continuing our discussion in the next sections on applications of property (T) to our problems, we make the following technical remark regarding finitely generated groups.

Remark 3.2.5. If Γ is a finitely generated discrete group with property (T), then there exist a compact (i.e., finite) subset K and an $\varepsilon > 0$ such that for every not trivial irreducible unitary representation (H, ρ) and for every $v \in H$ there exists $k \in K$ such that $\| \rho(k)v - v \| > \varepsilon \| v \|$. This is by (3.1.3). Once this is known to be true for one finite subset K, we claim that the same conclusion holds with every finite set S of *generators* of Γ. Indeed, if S is such a generating set, then every $w \in K$ is a word of length at most, say ℓ, in the elements of S. If (H, ρ) is a representation and $v \in H$ which satisfies $\| \rho(s)v - v \| < \varepsilon \| v \|$ for every $s \in S$, then $\| \rho(k)v - v \| < \ell\varepsilon \| v \|$ for every $k \in K$. This shows that S also does not have almost invariant vectors, if K does not.

3.3 Explicit construction of expanders using property (T)

We are now ready to apply property (T) to get explicit examples of expanders. This application is due to Margulis [M1] and was a breakthrough since it was the first explicit construction of expanders. Yet we first present a variant of it as presented in [AM]. Margulis' examples are presented in (3.3.5).

Proposition 3.3.1. *Let Γ be a finitely generated Kazhdan group (e.g.: $\Gamma = SL_n(\mathbb{Z})$, $n \geq 3$). Let $\mathcal{L}$ be a family of finite index normal subgroups of Γ and S a fixed finite symmetric (i.e., $S^{-1} = S$) set of generators for Γ. Then the family of $X(\Gamma/N, S)$, the Cayley graphs of the finite groups Γ/N, for $N \in \mathcal{L}$, with respect to S (considered as a set of generators of Γ/N) is a family of (n, k, c)-expanders for some $c > 0$, $k = |S|$ and $n = |\Gamma/N|$.*

The Cayley graph of a group G with respect to a subset S is the graph whose vertices are the elements of G, and $a \in G$ is connected to $as \in G$ for every $s \in S$.

Proof. Let $N \in \mathcal{L}$ and $H = L^2(\Gamma/N)$ the vector space of the complex valued functions on the finite set $V = \Gamma/N$, with the norm $\| f \|^2 = \sum_{x \in V} |f(x)|^2$. Let $H_0 = \{f \in H \mid \sum_{x \in V} f(x) = 0\}$. Then Γ acts on H by $(\gamma f)(x) = f(x\gamma)$, and $H = H_0 \oplus \mathbb{C}\chi_V$ as Γ-module. The action of Γ on V is transitive, so the only Γ-invariant functions on V are the constants $\mathbf{C}\chi_{\mathbf{V}}$. Thus H_0 does not contain the trivial representation and therefore, since Γ is Kazhdan, H_0 does not have almost invariant functions. This implies, in particular, that there are no almost invariant subsets, which is exactly the property of being an expander. To be more precise, there exists an $\varepsilon > 0$, dependent only on Γ and S but not on N, such that for every $f \in H_0$, $\| \gamma f - f \| > \varepsilon \| f \|$ for some $\gamma \in S$ (see (3.1.3) and (3.2.5)).

Let A be a subset of V of size a and B its complement of size $b = n - a$. Let

$$f(x) = \begin{cases} b & \text{if } x \in A \\ -a & \text{if } x \in B. \end{cases}$$

Then $f \in H_0$ and

$$\| f \|^2 = ab^2 + ba^2 = nab$$

while for every $\gamma \in S$, $\| \gamma f - f \|^2 = (b+a)^2 |E_\gamma(A,B)|$ where $E_\gamma(A,B) = \{x \in V \mid x \in A \text{ and } x\gamma \in B \text{ or } x \in B \text{ and } x\gamma \in A\}$. To summarize, there exists $\gamma \in S$ such that:

$$|\partial A| \geq \frac{1}{2}|E_\gamma(A,B)| = \frac{\| \gamma f - f \|^2}{2n^2} \geq \frac{\varepsilon^2 \| f \|^2}{2n^2} = \varepsilon^2 \frac{ab}{2n} = \frac{\varepsilon^2}{2} \cdot (1 - \frac{|A|}{n})|A|.$$

Thus $X(\Gamma/N, V)$ are expanders with $c \geq \frac{\varepsilon^2}{2}$. □

Example 3.3.2. Let $\Gamma_n = SL(n,\mathbb{Z}), n \geq 3$, and let

$$A_n = \begin{pmatrix} 1 & 1 & \vdots & 0_{2\times(n-2)} \\ 0 & 1 & \vdots & \\ \cdots & \cdots & \cdots & \cdots \\ 0_{(n-2)\times 2} & & \vdots & I_{n-2} \end{pmatrix}, B_n = \begin{pmatrix} 0 & 1 & & & \\ 0 & 0 & 1 & & \\ & & 0 & 1 & \\ & & & \ddots & \ddots \\ & & & \ddots & 1 \\ (-1)^{n-1} & & & & 0 \end{pmatrix}.$$

Then, it is well known that $S_n = \{A_n, B_n\}$ generates Γ_n. Fix n, and let $SL_n(p)$ be the group of $n \times n$ matrices of determinant 1 over the finite field of p elements. Then using the finite index normal subgroups $\Gamma_n(p) = \text{Ker}(SL_n(\mathbb{Z}) \to SL_n(p))$, we deduce that $X(SL_n(p), S_n) = X(n,p)$ is a family of expanders *when n is fixed* and p runs over all prime numbers.

We will see later on, in Chapter 4, that the same conclusion holds also for $n = 2$ but for a different reason (as $SL_2(\mathbb{Z})$ does not have property (T)). On the other hand:

Proposition 3.3.3. *([LW]) For a fixed prime p and n going to infinity, $X(n,p)$ is not a family of expanders.*

Proof. (Y. Luz) $SL_n(p)$ acts transitively on the nonzero vectors of $\mathbb{F}_p^n$. If the $X(n,p)$ would be expanders, then their quotient graphs $Z(n,p)$ would be expanders, where the vertices of $Z(n,p)$ are $\mathbb{F}_p^n \backslash \{0\}$ and $\alpha \in Z(n,p)$ is adjacent to $A_n^{\pm}(\alpha)$ and $B_n^{\pm}(\alpha)$. Let $Y_n \subseteq Z(n,p)$ be the subset $\{e_1, \ldots, e_{[n/2]}\}$ where $\{e_i\}$ is the standard basis of $\mathbb{F}_p^n$; then $A_n^{\pm}(Y_n) = Y_n$ while $|B_n^{\pm}(Y_n) \Delta Y_n| \leq \frac{5}{n}|Y_n|$. Hence $Z(n,p)$ and $X(n,p)$ are not expanders. □

A second proof based on Proposition 3.3.7 below is given in [LW].

It is not known, however, whether there exists a different set of generators $\mathcal{S}_n$ for $SL_n(p)$ (p fixed) for which $X(SL_n(p), \mathcal{S}_n)$ form a family of expanders (see §10.3).

Remark 3.3.4. The proof of (3.3.1) shows that the assumption of $N \in \mathcal{L}$ being normal was superfluous. N can be arbitrary finite index subgroup, in which case we take what is sometimes called the *Schreier graph* $X(\Gamma/N, S)$, i.e., the graph whose vertices are the left cosets of Γ modulo N and two cosets aN and bN are connected by $s \in S$ if $saN = bN$.

The proof of (3.3.1) also shows that we actually do not need the full power of property (T). We only need the non-trivial representations occurring in $L^2(\Gamma/N)$ to be bounded away from the trivial representation of Γ, in the unitary dual of Γ. In (3.1.11) we showed that the representations of $G = \mathbb{R}^2 \rtimes SL_2(\mathbb{R})$ that do not contain $\mathbb{R}^2$-invariant vectors are bounded away from the trivial representation. The group $\Gamma = \mathbb{Z}^2 \rtimes SL_2(\mathbb{Z})$ is a lattice in G, and hence if $R = \{\rho \in \hat{\Gamma} \mid \rho$ does not have $\mathbb{Z}^2$-invariant vector$\}$ then Γ has property $(T : R)$. This was the way Margulis gave the first examples of expanders, which are actually simpler than those given in (3.3.1):

Proposition 3.3.5. *Let m be a positive integer and $V_m = \mathbb{Z}/m\mathbb{Z} \times \mathbb{Z}/m\mathbb{Z}$. Define a graph on the set V_m by connecting every $(a,b) \in V_m$ to $\sigma_1(a,b) = (a+1,b), \sigma_2(a,b) = (a,b+1), \sigma_3(a,b) = (a,a+b)$ and $\sigma_4(a,b) = (-b,a)$. Then $\{V_m\}$ is a family of expanders.*

Proof. First, it is easy to see $\{\sigma_1,\dots,\sigma_4\}$ as elements of $\Gamma = \mathbb{Z}^2 \rtimes SL_2(\mathbb{Z})$ generate Γ. Secondly, Γ acts in a natural way on V_m by affine transformations and hence also on $L^2(V_m)$, and the only $\mathbb{Z}^2$-invariant functions are the constant functions since $\mathbb{Z}^2$ acts transitively via σ_1 and σ_2. The preceding remark implies now that V_m are indeed expanders. □

Margulis presented these graphs without estimating the constant c since this c is dependent, as we saw, on the ε in property (T), and the last is known to exist but with no estimate. Gaber and Galil [GG] changed slightly the generators and gave a very nice direct proof *including an estimation* of the constant c. We will not bring it here since by now there are better estimates, but the reader is encouraged to look at their paper. They also describe in detail how expanders lead to superconcentrators. Several authors (e.g., [An], [C1], [JM], [AM], [AGM]) elaborated on their work, either by changing the generators, or improving the proof or the construction to get better and better expanders and superconcentrators. We mention here only one negative result of Klawe [Kl1,2] which can be well understood from our perspective:

Proposition 3.3.6. *Given fixed k and c, if for every m we choose a set $S = \{\sigma_{1,m},\dots,\sigma_{k,m}\}$ of affine transformations of $W_m = \mathbb{Z}/m\mathbb{Z}$, then the graphs determined by them are not (m,k,c)-expanders for all but finitely many values of m.*

Proof. Let A_m be the group of affine transformations of $\mathbb{Z}/m\mathbb{Z}$, i.e., transformations of the form: $x \mapsto ax + b$ where $(a, m) = 1$ and $a, b \in \mathbb{Z}/m\mathbb{Z}$. This is a metabelian group, i.e., $A_m'' = \{1\}$. Thus the group generated by $\sigma_{1,m}, \ldots, \sigma_{k,m}$ is a quotient, through a homomorphism φ_m, of F, the free metabelian group on k generators $F = F_k/F_k''$, where F_k is the free group on k generators. The graph W_m is therefore a Schreier graph of F with respect to N_m where $N_m = \{g \in F \mid \varphi_m(g)(1) = 1\}$. The group F is solvable and hence amenable. The next general proposition form [LW] will finish the proof.

Proposition 3.3.7. *Let Γ be an amenable group generated by a finite set S. Let $\mathcal{L}$ be an infinite family of finite index subgroups. Then $\{X(\Gamma/N, S) \mid N \in \mathcal{L}\}$ is not a family of expanders.*

Proof. By (2.2.3) there is a finite subset $A \subseteq \Gamma$ such that $|A \triangle sA| < \varepsilon |A|$ for every $s \in S \cup S^{-1}$. For $N \in \mathcal{L}$, set a function $\varphi : G/N \to \mathbb{N}$, by: $\varphi(X) = |A \cap X|$ where $X = xN$ is a coset of N.

Then:

$$\sum_{X \in G/H} |(A \cap X) \triangle (sA \cap X)| + |(A \cap X) \triangle (s^{-1}A \cap X)|$$

$$\leq |A \triangle sA| + |A \triangle s^{-1}A| \leq 2\varepsilon |A|.$$

The L_1-norm of φ is: $\sum_{X \in G/H} \varphi(X) = |A|$. So: $|s\varphi - \varphi|_{L_1} \leq 2\varepsilon |\varphi|_{L_1}$. Thus φ is an «almost invariant function» in the L_1-norm.

We have to show that there exists «an almost invariant» subset. For $j \in \{1, 2, 3, \ldots\}$ define $B_j = \{X \in G/H \mid \varphi(X) \geq j\}$. Then it is easy to check that $\varphi = \sum_{j=1}^{\infty} \chi_{B_j}$ where χ_{B_j} is the characteristic function of B_j. Also we check that:

$$\sum_{j=1}^{\infty} |B_j \triangle sB_j| = \sum_{X \in G/H} |\varphi(X) - \varphi(s^{-1}X)| + |\varphi(X) - \varphi(sX)| \leq 2\varepsilon |A|. \tag{1}$$

For $s \in S \cup S^{-1}$, let $J_s = \{j \mid |B_j \triangle sB_j| > \varepsilon^{1/2} |S| |B_j|\}$. So:

$$|S| \varepsilon^{1/2} \sum_{j \in J_s} |B_j| \leq \sum_{j \in J_s} |B_j \triangle sB_j| \overset{(1)}{\leq} 2\varepsilon |A|.$$

Thus:

$$\sum_{j \in J_s} |B_j| < \frac{2\varepsilon^{1/2} |A|}{|S|}. \tag{2}$$

But

$$|A| = \sum_{j=1}^{\infty} |B_j|. \tag{3}$$

(The reader can convince himself that this is the case by drawing a «graph» of the function φ and observing that (3) is just a «Fubini Theorem»). (2) and (3) give:

$$\sum_{j \in J_s} |B_j| < \frac{2\varepsilon^{1/2}}{|S|} (\sum_{j=1}^{\infty} |B_j|)$$

and so:

$$\sum_{s \in S \cup S^{-1}} \sum_{j \in J_s} |B_j| < 2 \cdot 2\varepsilon^{1/2} (\sum_{j=1}^{\infty} |B_j|).$$

This means that there exists j_0 such that $j_0 \notin J_s$ for $s \in S \cup S^{-1}$, i.e., $|B_{j_0} \triangle s B_{j_0}| < |S| \varepsilon^{1/2} |B_{j_0}|$ for every $s \in S \cup S^{-1}$.

This provides us with the desired «almost invariant set». □

The reader may observe that, just like in (3.1.5), the essential part of the proof is to show that if there are «almost invariant functions» then there are also «almost invariant subsets». (Compare also (4.2.4).)

3.4 Solution of the Ruziewicz problem for S^n, $n \geq 4$

Finitely generated Kazhdan groups can be used also to answer the Ruziewicz problem. Let's recall that Ruziewicz asked whether every finitely additive rotation invariant measure on S^n of total measure one, defined on all Lebesgue measurable sets, is equal to the Lebesgue measure. We saw that for S^1 the answer is negative. For $n \geq 2$, we have proved that such a measure is absolutely continuous with respect to Lebesgue measure. It can therefore be used in a standard way to define an integral on $L^\infty(S^n)$. So Ruziewicz's problem amounts to asking whether the Lebesgue integral is the only $SO(n+1)$ invariant mean on $L^\infty(S^n)$.

Proposition 3.4.1. *Let Γ be a finitely generated subgroup of $SO(n+1)$. Γ acts on S^n and therefore on $L^2(S^n)$. Let ρ' be this unitary representation and ρ be the restriction of ρ' to the subspace $L_0^2(S^n) = \{f \in L^2(S^n) \mid \int_{S^n} f d\lambda = 0\}$. If ρ does not weakly contain the trivial representation, then the Lebesgue integral is the unique invariant mean on $L^\infty(S^n)$.*

Proof. Let S be a finite set of generators for Γ. Let $X = L^\infty(S^n)$ and $m \in X^*$ an invariant mean. Now $X = (L^1)^*$ and hence $m \in (L^1)^{**}$. L^1 is a weak* dense in $(L^1)^{**}$ and since m is nonnegative on nonnegative functions and $m(\chi_{S^n}) = 1$, we can find a net $\{f_\alpha\} \in L^1$ with $\int f_\alpha d\lambda = 1$, $f_\alpha \geq 0$ and $\lim f_\alpha = m$ in the weak* topology. Moreover, for every $\gamma \in S$, $\gamma m = m$ so $\lim(\gamma f_\alpha - f_\alpha) = 0$ weakly in L^1 (since it does so in the weak* topology of $(L^1)^{**}$). The weak and the strong closures of convex sets are the same in L^1 (cf. [Ru2, 3.12]), so we can take convex combinations of the $f'_\alpha - s$ to get a new net $\{g_\beta\}$ for which $\lim(\gamma g_\beta - g_\beta) = 0$ strongly in L^1 for every $\gamma \in S$ and also $\int g_\beta d\lambda = 1$, $g_\beta \geq 0$ and $\lim g_\beta = m$ weak* in $(L^1)^{**}$. Let $F_\beta = \sqrt{g_\beta}$; then $F_\beta \in L^2(S^n)$ and $\| \gamma F_\beta - F_\beta \|_{L^2}^2 \leq \| \gamma g_\beta - g_\beta \|_{L^1} \to 0$ for each $\gamma \in S$. Thus we have $\| F_\beta \|_{L^2} = 1$ and $\lim \| \gamma F_\beta - F_\beta \|_{L^2} = 0$ so they are almost invariant in $L^2(S^n)$. By our assumption on Γ, we deduce that F_β converges to χ_{S^n} (note that $L^2(S^n) = \mathbb{R}\chi_{S^n} \oplus L_0^2(S^n)$ and the projection of F_β into the right-hand side factor must go to zero by our assumption while the projection to $\mathbb{R}\chi_{S^n}$ is χ_{S^n}). We deduce therefore that (1 stands for $1\chi_{S^n}$):

$$\begin{aligned} \| g_\beta - 1 \|_{L^1} &= \int |F_\beta^2 - 1| d\lambda = \int |F_\beta - 1||F_\beta + 1| d\lambda \\ &\leq \| F_\beta - 1 \|_{L^2} \| F_\beta + 1 \|_{L^2} \leq 2 \| F_\beta - 1 \|_{L^2} . \end{aligned}$$

Hence g_β converges strongly in L^1 to χ_{S^n}, but also $\lim g_\beta = m$ weak* in $(L^1)^{**}$. This shows that m is just integration against χ_{S^n}, i.e., it is no more than the Lebesgue integral λ. □

The above-mentioned result is due to Rosenblatt [Ro1] and Del-Junco-Rosenblatt [DJR] (while the above simple proof is due to Losert and Rindler [LR]). From it Margulis [M2] and Sullivan [Su] deduced:

Theorem 3.4.2. *Lebesgue measure is the only finitely additive measure defined on all Lebesgue measurable subsets of S^n $(n \geq 4)$ invariant under the action of $SO(n+1)$ and of total measure* 1.

Proof. Let m be such a measure. We have to prove $m = \lambda$ where λ is the Lebesgue measure. By (2.2.12) m is absolutely continuous with respect to λ. By the standard way to define an integral using a given measure, m defines an invariant mean on $L^\infty(S^n)$. To prove that $m = \lambda$ it is sufficient to find a finitely generated group Γ which has property (T) and can be densely embedded in $SO(n)$, $n \geq 5$. Once this is done, Γ does not have any invariant function in $L_0^2(S^n)$ (since its action on S^n is ergodic) and therefore its representation on $L_0^2(S^n)$ cannot contain ρ_0 weakly. By (3.4.1) this implies $\lambda = m$. So the next proposition will finish the proof.

Proposition 3.4.3. *For $n \geq 5$, $SO(n)$ has a finitely generated dense subgroup with property (T).*

Proof. Let Γ be an arithmetic group in $SO(n-2,2)$ as constructed in Example (3.2.2C). Then Γ is constructed as a subgroup of $SO(n-2,2) \times SO(n)$, and since $n-2 \geq 3$, it has property (T) by (3.2.4C). The projection of Γ into $SO(n)$, $\pi_2(\Gamma)$, is dense. Otherwise, its closure G is a proper subgroup whose complexification $G_{\mathbb{C}}$ in the complexification $SO(n,\mathbb{C})$ of $SO(n)$ is a proper algebraic subgroup containing $\pi_2(\Gamma)$. But applying the Galois automorphism σ (see (3.2.2C)) we get $\pi_1(\Gamma)$ (where π_1 is the projection of Γ into $SO(n-2,2)$) as a lattice in $SO(n-2,2)$. By Borel's Density Theorem ([Zi1]), $\pi_1(\Gamma)$ is Zariski dense in the complexification of $SO(n-2,2)$, i.e., in $\underset{\sim}{G}(f)$. This implies that $\pi_2(\Gamma) = {}^{\sigma}\pi_1(\Gamma)$ is Zariski dense in ${}^{\sigma}\underset{\sim}{G}(\mathbb{C})$. It is therefore impossible that it is contained in a proper algebraic subgroup of $SO(n,\mathbb{C}) = {}^{\sigma}\underset{\sim}{G}(\mathbb{C})$. □

The above proof follows Sullivan [Su]. Margulis [M2] used a different lattice: Let $\Gamma = SO(n,\mathbb{Z}[\frac{1}{5}])$. Then it is a lattice in $SO(n,\mathbb{Q}_5) \times SO(n)$ by (3.2.2B) and it has property (T) if $n \geq 5$ (see (3.2.4B)). Its projection to $SO(n)$ is the desired group. Margulis also went ahead to prove:

Theorem 3.4.4. *Let G be any real compact simple Lie group which is not locally isomorphic to $SO(n), n = 2, 3$ or 4. Then G has a finitely generated dense Kazhdan subgroup.*

The theorem is of interest since in the proof of 3.4.1, there was nothing special about $SO(n+1)$ and S^n. In fact, one can prove (see [Wa, Theorem 10.11]):

Theorem 3.4.5. *Suppose a finitely generated Kazhdan group Γ acts on X in a way that is measure-preserving and ergodic, where $(X, \mathcal{A}, m)$ is a non-atomic measure space and $m(X) = 1$. Then the m-integral is the unique G-invariant mean on $L^{\infty}(X)$.*

Theorems (3.4.4) and (3.4.5) together yield

Corollary 3.4.6. *Let G be a simple compact real Lie group which is not locally isomorphic to $SO(n)$, $n = 2, 3, 4$. Then the Haar measure is the only G-invariant finitely additive measure defined on the Haar measurable sets and of total measure one.*

In this Corollary $SO(2)$ should really be excluded, as we show in (2.2.11). On the other hand, the uniqueness of the Haar measure holds for $SO(3)$ and $SO(4)$. Indeed, what one needs is not necessarily a group with property (T) to be dense in G. It suffices (see (3.4.1)) to have a finitely generated group Γ in G whose representation on $L_0^2(G)$ does not weakly contain the trivial representation. In Chapter 7 we will show that such Γ's exist for $G = SO(3)$ or $SO(4)$. This will prove Corollary 3.4.6 for these groups as well as answer affirmatively the Ruziewicz problem for S^n, $n = 2$ and 3.

Still, in Theorem (3.4.4) the cases $n = 2, 3, 4$ are exceptional. This is clear for $SO(2)$ since it is abelian. The rest of this section is devoted to prove this fact for $SO(3)$. From this, the case $SO(4)$ follows since $SO(4)$ is locally isomorphic to $SO(3) \times SO(3)$. Instead of proving that $SO(3)$ does not contain an infinite finitely generated group with property (T), it suffices to prove it for its double cover $SU(2)$. We, in fact, prove a stronger result. The following theorem is due to Zimmer [Zi2]. We bring a somewhat different proof. Another proof can be found in [HV].

Theorem 3.4.7. *If Γ is a finitely generated Kazhdan subgroup of $SL_2(\mathbb{C})$, then Γ is finite. In particular, $SU(2)$ (and hence also $SO(n), n = 2, 3, 4$) does not have a dense countable Kazhdan subgroup.*

The proof will be divided up to several propositions. Some of them are of independent interest. We are actually using in the proof some notions which are presented in more detail in Chapters 5 and 6. Some of the readers might prefer to skip the proof in the first reading.

Let Γ be a finitely generated subgroup at $SL_n(\mathbb{C})$. So $\Gamma \subset SL_n(A)$ for some finitely generated subring A of $\mathbb{C}$ (take A to be the subring generated by the entries of a finite set of generators of Γ). A homomorphism φ from A to $\tilde{\mathbb{Q}}$ – the algebraic closure of $\mathbb{Q}$ – is called a *specialization* of A. By Hilbert's Nullstellensatz Theorem, these homomorphisms from A to $\tilde{\mathbb{Q}}$ separate the points of A. Every such homomorphism induces a homomorphism, denoted also φ from $SL_n(A)$ to $SL_n(\tilde{\mathbb{Q}})$. Moreover, the image indeed lies in $SL_n(k)$ for some finite extension k of $\mathbb{Q}$.

Proposition 3.4.8. *If Γ is infinite and not abelian by finite, then for some specialization φ, $\varphi(\Gamma)$ is also infinite.*

Proof. If M is a finite subgroup of $SL_n(k) \subseteq SL_n(\mathbb{C})$, then by a theorem of Jordan (cf. [Ra1, 8.29]), there exists in M an abelian normal subgroup of index at most $j(n)$. Let K be the intersection of all subgroups of Γ of index at most $j(n)$. As Γ is finitely generated, K is of finite index in Γ. If for every φ, $\varphi(\Gamma)$ is finite, then $\varphi([K, K]) = \{1\}$ for every φ. Since the specializations separate the points of Γ, this yields that $[K, K] = \{1\}$ and Γ is abelian by finite, a contradiction. $\square$

Proposition 3.4.9. *Let Γ be a subgroup of $SL_n(k)$ for some number field k. Assume that for every embedding of k into a local field k_ν, the image of Γ in $SL_n(k_\nu)$ is precompact. Then Γ is finite.*

Proof. Let $\mathcal{P}$ be the set of all valuations of k and $\mathbb{A}_k = \prod'_{\nu \in \mathcal{P}} k_\nu$ the adèle ring of k (see Chapter 6 below). Then $SL_n(k)$ is a discrete subgroup of $SL_n(\mathbb{A}_k)$. So, if the image of Γ in $SL_n(k_\nu)$ is precompact for every ν, the image of Γ in $SL_n(\mathbb{A}_k)$ is precompact. So it is precompact and discrete. Hence, Γ is finite. $\square$

The following proposition is due to Watatani [Wt] (see also Alperin [Alp] and more generally Adams and Spatzier [AS]). We give here a different proof.

Proposition 3.4.10. *Let Γ be a group with property (T) acting on a locally finite tree X without inversions. Then Γ fixes a vertex of X.*

Proof. Let Y be a minimal Γ-invariant subtree of X. Then we can assume that $\Gamma \subseteq \mathrm{Aut}(Y)$. As Γ is finitely generated, Y/Γ is finite (see Bass [Bss2]). $\mathrm{Aut}(Y)$ is a locally compact group with the usual topology. Let H be the closure of Γ. Since Γ has property (T), so does H. In particular H is unimodular (since the commutator quotient is compact). By Bass-Kulkarni ([BK, Theorem 4.7]) H has a uniform Y-lattice, i.e., a discrete subgroup L such that Y/L is finite. Such L is either finite or contains a non-trivial free subgroup on ℓ generators ($\ell \geq 1$) of finite index. The latter case is impossible since L also has property (T) as L is a cocompact lattice in H (3.1.8(ii)). Thus L is finite and H is compact. This implies that H has a fixed vertex and hence also Γ. □

Proposition 3.4.11. *Let Γ be a group with property (T), and F a local field. For every homomorphism of Γ into $SL_2(F)$, the image of Γ is precompact.*

Proof. If F is a local non-archimedian field, then $SL_2(F)$ acts on its Bruhat-Tits tree (cf. Chapter 5 or [S1]), where the stabilizer of a vertex is a compact subgroup. By (3.4.10) it fixes a vertex. Hence the image of Γ is precompact.

If F is archimedian then $F = \mathbb{R}$ or $\mathbb{C}$. It suffices to consider $\mathbb{C}$. So assume $\Gamma \to SL_2(\mathbb{C})$. If it is not Zariski dense then it must have a solvable subgroup of finite index, which is impossible unless Γ is finite. So, Γ is Zariski dense. Hence its topological closure is either discrete or is a closed Lie subgroup whose real Lie algebra spans over $\mathbb{C}$ the Lie algebra $\mathcal{S}\ell_2(\mathbb{C})$. The first case, i.e., Γ discrete, is impossible unless Γ is finite: This is because if Γ is a lattice then it does not have property (T). A quick way to see that is to use (3.1.8) and to show that *one* lattice is not Kazhdan. Indeed, every finitely generated discrete subgroup of $SL_2(\mathbb{C})$ which is not cocompact contains a finite index subgroup which is mapped onto $\mathbb{Z}$ (see Epstein [Ep] for an elegant short proof using Euler characteristic; see also [Lu1]). So no discrete subgroup of $SL_2(\mathbb{C})$ has property (T) unless it is finite. This also implies that $SL_2(\mathbb{C})$ does not have (T) and so Γ is not dense there.

The other possibility is that a finite index subgroup of Γ is dense in $SL_2(\mathbb{R})$ or in $SU(2)$. The first alternative is impossible since $SL_2(\mathbb{R})$ does not have property (T). The second one is the desired one. It means Γ is precompact. □

Remark 3.4.12. Zimmer [Zi2] gives a simple lemma that asserts that if $\varphi : \Gamma \to G$ is a homomorphism, Γ is Kazhdan, G is not Kazhdan and G satisfies the Howe-Moore property ([HM]), i.e., the matrix coefficients of unitary non-trivial irreducible representations of G vanish at infinity, then $\varphi(\Gamma)$ is precompact. Since $G = SL_2(F)$ satisfies all these assumptions, Proposition 3.4.10 follows with a more uniform proof.

De la Harpe and Valette in [HV] prove this proposition while using Watatani's result for the non-archimedian case and by proving an analogue of it for the hyperbolic case. Adams and Spatzier [AS] have proved an even stronger «fixed point property» for Kazhdan groups.

Proof of Theorem 3.4.7. If $\Gamma \subset SL_2(\mathbb{C})$ is a finitely generated infinite Kazhdan subgroup then it is not abelian by finite, and so we can assume by (3.4.8) that Γ is in $SL_2(k)$ for some number field k. By (3.4.11) the assumptions of (3.4.9) are satisfied and so Γ is finite. □

3.5 Notes

1. Property (T) was defined by Kazhdan in [Ka], who also proved Theorem 3.1.15 above. To be more precise, Kazhdan announced it only for rank ≥ 3, but it was very quickly realized by Wang [Wan1], Delaroche-Kirillov [DKi] and Vaserstein [Vs] that the same holds for rank ≥ 2 and over every local field. Some of the $\mathbb{R}$-rank one real semi-simple Lie groups have property (T): $Sp(n, 1)$ and $\mathbf{F_4}$, while the rest of them $SO(n, 1)$ and $SU(n, 1)$ do not. On the other hand over non-archimedian local fields, no rank-one semi-simple group has property (T). The book of P. de la Harpe and A. Valette [HV] contains all this information and much more. Other clear expositions of property (T) are in Zimmer's book [Zi1] and Margulis' [M7].
2. Kazhdan groups play the role of the other extreme to amenable groups. This can be seen from what we have proved, but better illustrations are the following two stronger results (which should be compared with our §2.2 and with Theorem (3.4.5)):

   Theorem. *Let Γ be an infinite countable group. Then:*
   (a) Γ is amenable if and only if for every ergodic, measure-preserving action of Γ on a non-atomic measure space $(X, \mathcal{A}, m)$ with $m(X) = 1$, $L^\infty(X, m)$ has more than one Γ-invariant mean.
   (b) Γ is Kazhdan if and only if for every action as in (a), the m-integral is the only Γ-invariant mean.

   The above theorem is included in the result of Rosenblatt [Ro1], A. Connes and B. Weiss [CW] and K. Schmidt [Sc2].
3. Margulis [M1] gave the examples of expanders given here in (3.3.5), but, as was mentioned there, without estimating the expansion coefficient c. Gabber and Galil [GG] took different generators for $SL_2(\mathbb{Z}) \ltimes \mathbb{Z}^2$ and got slightly different expanders for which they estimated the constants. Their expanders (plus some ideas in the construction of super-concentrators) led to $(n, \ell = 261.78)$-superconcentrators. A series of papers elaborated on [GG] either by changing the generators or by improving some steps of the constructions: F. Chung reduced it to $\ell = 261.5$, Jimbo and Maruoka [JM] to $\ell = 248$, Alon-Galil-Milman got $\ell = 157.35$ and then improved it in [AGM] to $\ell = 122.74$.

Finally, the «Ramanujan graphs» constructed by Lubotzky-Phillips-Sarnak [LPS2] give rise to $(n, 64)$ superconcentrators (as explained in [Bc]). This is close but not yet as good as the $(n, 36)$-superconcentrators which are known to exist by random consideration.

4. It is interesting to mention here the paper of N.J. Kalton and J.W. Roberts [KR], where a problem about measures is solved using expanders and concentrators.
5. In [CLR] the authors related a property of approximation of compact operators on $L^2(G)$ by sums of translations to the question whether ρ_0 is contained weakly in $L_0^2(G)$. They were able, therefore, to apply property (T) to this problem in a similar way to §3.4.

   Similarly, Rosenblatt [Ro2] used the fact that ρ_0 is not weakly contained in $L_0^2(G)$ for any simple compact real Lie group (by (3.4.4) and also (7.2.1) below) to conclude that every left translation invariant linear form on $L_p(G)$, $1 < p < \infty$, is continuous. This extends earlier results of Musters and Schmidt and of Bourgain.
6. Property $(T : R)$ introduced in (3.1.10) seems to be of some independent interest; see [LZ] for more details. It is shown there, for example, that if Γ is an irreducible lattice in a product $G_1 \times G_2$ where G_1 is Kazhdan (but not necessarily G_2), then Γ has property $(T : F)$ where F is the family of finite-dimensional unitary representations of Γ. In particular, the finite quotients of Γ form a family of expanders just as in (3.3.1), even though Γ does not have property (T). This does not give any interesting expanders from a computer science point of view, but it does carry some geometric information (see §4.2).
7. The ε mentioned in Definition (3.1.3) is sometimes referred to as the Kazdhan constant of G with respect to K. In our Definition (3.1.3) we required that ε «works» only for irreducible non-trivial representation. It implies, however, that there exists an ε' which works for every unitary representation which does not contain the trivial one. But ε and ε' might be different as illustrated in [BDH].

4 The Laplacian and its Eigenvalues

4.0 Introduction

Let M be a connected compact smooth Riemannian manifold, and let $\Delta = -\operatorname{div}(\operatorname{grad})$ its Laplacian operator of $L^2(M)$. Its eigenvalues $\lambda_0 = 0 < \lambda_1(M) \leq \lambda_2(M) \leq \cdots$ form a discrete subset (with multiplicities) of $\mathbb{R}_+$.

As was exhibited in various ways, this spectrum reflects a lot of the geometry of M. We will be interested mainly in $\lambda_1 = \lambda_1(M)$ – the first positive eigenvalue. By results of Cheeger and Buser, λ_1 reflects the «isoperimetric constant of M».

In an analogous way one defines the Laplacian Δ of a graph X. If X is k-regular, Δ is simply equal to $kI - \delta$, where δ is the adjacency matrix of X. Here also we denote $\lambda_1 = \lambda_1(X)$. By analogous results due to Dodziuk, Tanner, Alon and Milman, $\lambda_1 = \lambda_1(X)$ reflects the isoperimetric constant of X, i.e., the expansion factor of it. It turns out that the question of finding a family of k-regular expanders is equivalent to the question of finding graphs for which λ_1 is uniformly bounded away from zero.

This connection between manifolds and graphs gives more than just an analogy. A deep theorem of Selberg asserts that $\lambda_1(\Gamma(m)\backslash\mathbb{H}) \geq \frac{3}{16}$ where $\mathbb{H}$ is the upper half-plane with the hyperbolic metric and $\Gamma(m) = \ker(SL_2(\mathbb{Z}) \to SL_2(\mathbb{Z}/m\mathbb{Z}))$ are congruence subgroups of the modular group. (He conjectured that in fact $\lambda_1(\Gamma(m)\backslash\mathbb{H}) \geq \frac{1}{4}$). The Cayley graphs of $SL_2(\mathbb{Z})/\Gamma(m)$ with respect to appropriate generators of $SL_2(\mathbb{Z})$ can be presented as «discrete approximations» of the manifolds $\Gamma(m)\backslash\mathbb{H}$, and λ_1 of these graphs can be evaluated using the above-mentioned result of Selberg. This leads to an explicit construction of expanders with an explicit expansion constant.

For a k-regular graph X, $\Delta = kI - \delta$, so we can equivalently look at the eigenvalues of δ. They control the speed of convergence of the random walk on X to the uniform distribution. To ensure fast convergence one wants the non-trivial eigenvalues of δ to be as small as possible in absolute values. It turns out that for an infinite family of k-regular graphs, the best possible bound we can hope for is $2\sqrt{k-1}$. This leads to the important notion of a Ramanujan graph, i.e., a k-regular graph X for which the eigenvalues λ of δ are either $\pm k$ or $|\lambda| \leq 2\sqrt{k-1}$. Such graphs are very good expanders. Selberg Theorem does not give Ramanujan graphs – but it is a hint in the right direction, as will be evident in the next chapters.

4.1 The geometric Laplacian

In this section we survey briefly the geometric Laplacian and some basic results on its eigenvalues. Let M be a connected Riemannian manifold of dimension n (which we always assume to be without a boundary), i.e., M is a differentiable manifold with a smooth choice of a positive definite inner product $(\,:\,,\,:\,)_m$ on the tangent space T_m for every point m of M. Let $V = T_m^*$ be the dual of $T_m, \Lambda(V)$ the exterior algebra of V, and $\Lambda_k(V)$ the homogeneous elements of degree k. $\Lambda_k^*(M) = \bigcup_{m \in M} \Lambda_k(T_m^*)$ is the exterior k-bundle over M. A C^∞-map from M to $\Lambda_k^*(M)$ (resp: $\Lambda^*(M)$) whose composition with the canonical projection is the identity map of M is called a k-form on M (resp: a form on M). The space of k-forms will be denoted $E^k(M)$, and $E^*(M) = \sum_{k=0}^n E^k(M)$. $E^0(M)$ can be identified with $C^\infty(M)$. The inner product on T_m^* gives a canonical identification of T_m^* and T_m. This inner product can be extended to $\Lambda(T_m^*)$ and by integration over the manifold also to an inner product on $E^*(M)$ (see [War]). More precisely, if $\alpha, \beta \in E^k(M)$ then

$$\langle \alpha, \beta \rangle = \int_M \alpha \Lambda^* \beta$$

where * is the star operation $*: E^k(M) \to E^{n-k}(M)$.) The operator $d : C^\infty(M) \to E^1(M)$ given by $(df)(X) = X(f)$ where X is a vector field on M can be also extended in a natural way to a map from $E^k(M) \to E^{k+1}(M)$ and therefore defines $d : E^*(M) \to E^*(M)$. Let δ be its adjoint; in fact, δ takes k-forms to $(k-1)$-forms and on k-forms it is equal to $(-1)^{n(k+1)+1} {*d*}$ where *, as before, is the star operation. We can now define the Laplacian (or the Laplace-Beltrami operator) as $\Delta = \delta d + d\delta$.

We gave the general definition for future reference (see (10.8.8)), but we will be interested here merely in its action on $C^\infty(M) = E^0(M)$. Here $\delta = 0$ so $\Delta = \delta d = -\operatorname{div} \circ d$ (or with the identification of T_m^* with T_m and so $df = \operatorname{grad}(f)$, we can write $\Delta = -\operatorname{div} \circ \operatorname{grad}$, which is the more standard definition). Δ satisfies the condition that for any two functions $f, g \in C^\infty(M)$, $\int_M f \Delta g = \int_M \langle \operatorname{grad} f, \operatorname{grad} g \rangle$. This can be also taken as a definition of Δ. To be more specific, if $f \in C^\infty(M)$ then Δ on $C^\infty(M)$ is a second-order differential operator which in local coordinates is given by:

$$\Delta f = \frac{-1}{\sqrt{h}} \sum_{i=1}^n \frac{\partial}{\partial x_i} \left(\sum_{j=1}^n h^{ij} \sqrt{|h|} \frac{\partial f}{\partial x_j} \right)$$

where (h_{ij}) is the matrix defining the Riemannian metric, namely $h_{i,j} = \left(\frac{\partial}{\partial x_i}, \frac{\partial}{\partial x_j} \right)$, (h^{ij}) is its inverse and $|h| = |\det(h_{ij})|$.

For example, for $\mathbb{R}^n$ with the standard metric

$$\Delta f = -\sum_{i=1}^n \frac{\partial^2 f}{\partial x_i^2}$$

while for the upper half-plane $\mathbb{H} = \{x + iy \mid x, y \in \mathbb{R}, y > 0\}$ with the metric $ds^2 = \frac{1}{y^2}(dx^2 + dy^2)$, one calculates:

$$\Delta f = -y^2 \left(\frac{\partial^2}{\partial x^2} + \frac{\partial^2}{\partial y^2} \right).$$

A more intuitive description of Δ is given by the following formula:

$$\lim_{r\to 0}\frac{1}{r^2}\left(\frac{\int_{S_r} f}{\operatorname{vol}(S_r)} - f(p)\right) = \frac{1}{2n}\Delta f(p)$$

where M is a Riemannian manifold of dimension n, $p \in M, r > 0$ and S_r is the sphere of radius r around p.

All these are equivalent forms of the Laplacian operator. We do not spend the time to show their equivalence, as we are mainly interested in its eigenvalues (in fact, only in the smallest positive one). This eigenvalue can be effectively studied using the Rayleigh quotient as we now explain.

If M is a manifold without a boundary, then Green's Theorem says that $\int_M g\Delta f = \int_M (\Delta g) f$, which shows that Δ is self-adjoint. It is also positive since $\int_M f(\Delta f) = \int_M \|\operatorname{grad} f\|^2 \geq 0$, and so its eigenvalues are real and nonnegative. The harmonic functions are the eigenfunctions with respect to the zero eigenvalue. In case M is compact, the only harmonic functions are the constants. In addition, there exists an orthonormal basis $\{\varphi_i\}_{i=0}^{\infty}$ of $L^2(M)$ consisting of eigenfunctions of Δ, i.e., $\Delta\varphi_i = \lambda_i\varphi_i$ (the φ_i are in $C^\infty(M)$) and $\int \phi_i = 0$ for $i \geq 1$. The sequence of eigenvalues $\{\lambda_i\}$ is a discrete subset of $\mathbb{R}_+$. We can order them so that $\lambda_0 = 0 < \lambda_1 \leq \lambda_2 \leq \cdots$.
There might be a repetition, but each value can appear only finitely many times.

Let $\lambda_1(M) = \lambda_1$ be the smallest positive eigenvalue. For our application $\lambda_1(M)$ is in fact the important notion rather than the Laplacian. If $f \in C^\infty(M)$ and $f \in L^2(M)$, then $f = \sum a_i\varphi_i$ where $a_i = \langle\varphi_i, f\rangle = \int_M \varphi_i f dm$. Thus:

$$\langle df, df\rangle = \langle\Delta f, f\rangle = \sum \lambda_i a_i^2 \geq \lambda_1 \sum (a_i^2) = \lambda_1 \|f\|^2.$$

We thus got an intrinsic characterization of $\lambda_1(M)$ independent of Δ.

Proposition 4.1.1. *Let M be a manifold (always without boundary). Then:*

$$\lambda_1(M) = \inf\left\{\frac{\int_M \|df\|^2}{\int_M |f|^2} \Big| f \in C^\infty(M) \text{ and } \int_M f = 0\right\}.$$

Definition 4.1.2. *Let M be an n-dimensional Riemannian compact manifold (or of finite volume). Let E be a compact $(n-1)$-dimensional submanifold which divides M into two disjoint submanifolds A and B. Let $\mu(E)$ be the area of E and $\lambda(A)$ and $\lambda(B)$ the volumes of A and B, respectively. The* **Cheeger constant** *$h(M)$ of M is defined as $\inf_E \frac{\mu(E)}{\min(\lambda(A),\lambda(B))}$ where E runs over all the possibilities for E as above.*

The following fundamental result is due to Cheeger [Cr].

Theorem 4.1.3. (Cheeger's inequality) $\lambda_1(M) \geq \frac{h^2(M)}{4}$.

Let M be an n-dimensional compact manifold and $R(M)$ its Ricci curvature, e.g., $R(\mathbb{R}^n) = 0$ and $R(\mathbb{H}) = -1$ (see [Ber], [BGM] and [Ch] for the general definition). Then Buser [Bu2] proved a converse to Cheeger's Theorem which says:

Theorem 4.1.4. *If* $R(M) \geq -(n-1)a^2$ *for some* $a \geq 0$ *where* $n = \dim M$*, then* $\lambda_1 \leq 2a(n-1)h(M) + 10h^2(M)$.

So for hyperbolic surfaces (quotients of $\mathbb{H}$ by lattices of $SL_2(\mathbb{R})$) we have: $h(M) \geq \frac{1}{10}\left(\sqrt{10\lambda_1 + 1} - 1\right)$ (as we take $n = 2$ and $a = 1$).

4.2 The combinatorial Laplacian

In analogy to the geometric Laplacian, one can define a combinatorial Laplacian for graphs (and, in fact, also for general CW-complexes).

Let $X = (V, E)$ be a finite graph. Fix some arbitrary orientation on the edges of the graph. For $e \in E$ we denote by e^- its origin and e^+ its target. We can think of e as a «tangent vector» to X at the point e^- pointing toward e^+. $L^2(E)$ can be therefore thought as the 1-forms of X, i.e., every $w \in L^2(E)$ associates a number $w(e)$ with every tangent vector. The operator $d : L^2(V) \to L^2(E)$ is defined as $df(e) = f(e^+) - f(e^-)$. If $|V| = n$ and $|E| = m$ then the matrix of d with respect to the standard bases of $L^2(V)$ and $L^2(E)$, respectively, is an $m \times n$ matrix D indexed by the pairs (e, v) where $v \in V$ and $e \in E$ (after fixing some order on V and E) such that

$$D_{e,v} = \begin{cases} 1 & \text{if } v = e^+ \\ -1 & \text{if } v = e^- \\ 0 & \text{otherwise.} \end{cases}$$

Let D^* be the transpose of D. Then D^*D is an $n \times n$ matrix.

Definition 4.2.1. $\Delta = D^*D$ *is called the Laplacian operator of the graph* X.

It is interesting to observe that while D depends on the orientation we put on X, Δ does not.

It follows from the definition that if $f, g \in L^2(V)$ then $\langle f, \Delta g\rangle = \langle df, dg\rangle$. In fact, this property characterizes Δ.

The next proposition gives a different interpretation of the Laplacian:

Proposition 4.2.2. *Let* $X = (V, E)$ *be a graph on* n *vertices, and* $\delta = \delta_X$ *its adjacency matrix, i.e., the* $n \times n$ *matrix indexed by the pairs* $(u, v) \in V \times V$ *and* $\delta_{u,v}$ *is equal to the number of edges between* u *and* v.

Let S *be the diagonal* $n \times n$ *matrix indexed by* $V \times V$*, where* $S_{v,v} = d(v)$ *is the degree of the vertex* v *and* $S_{u,v} = 0$ *if* $u \neq v$*. Then* $\Delta = S - \delta$*. In particular, if* X *is* k*-regular, then* $\Delta = kI - \delta$.

Remark. As an operator on $L^2(X)$ (i.e., $L^2(V)$), Δ is an averaging operator: if $f \in L^2(V)$ then

$$\Delta f(x) = \sum_{y \in V} \delta_{xy}(f(x) - f(y)) = d(x)f(x) - \sum_{y \in V} \delta_{xy} f(y).$$

(Note that usually δ_{xy} is either 0 or 1.) This formula serves as a definition for the Laplacian for every locally finite graph.

Proof. By the above it suffices to prove that $\langle f, (S - \delta)g \rangle = \langle df, dg \rangle$. It suffices, in fact, to prove it for real functions.

$$\begin{aligned}\langle f, (S - \delta)g \rangle &= \sum_{x \in V} f(x) \left[d(x)g(x) - \sum_{y \in V} \delta_{xy} g(y) \right] \\ &= \sum_{x \in V} d(x)f(x)g(x) - \sum_{x,y \in V} \delta_{xy} f(x)g(y).\end{aligned}$$

On the other hand:

$$\begin{aligned}\langle df, dg \rangle &= \sum_{e \in E} (df(e)) \cdot (dg(e)) \\ &= \sum_{e \in E} (f(e^+) - f(e^-))(g(e^+) - g(e^-)) \\ &= \sum_{e \in E} (f(e^+)g(e^+) + f(e^-)g(e^-)) - \sum_{e \in E} (f(e^+)g(e^-) + f(e^-)g(e^+)) \\ &= \sum_{x \in V} d(x)f(x)g(x) - \sum_{x,y \in V} \delta_{xy} f(x)g(y).\end{aligned}$$ □

The proposition shows that Δ is self-adjoint and positive, so its eigenvalues are real and nonnegative. If X is finite, then zero is always an eigenvalue with the constant function as an eigenfunction (so the only «harmonic" functions are the constants). It is a simple eigenvalue if and only if X is connected (cf. [Bg]), which we will assume throughout. Let $\lambda_1(X)$ be the smallest positive eigenvalue. Then in analogy to (4.1.1) we have:

Proposition 4.2.3.

$$\begin{aligned}\lambda_1(X) &= \inf \left\{ \frac{\sum_{e \in E} |df(e)|^2}{\sum_{x \in V} |f(x)|^2} \,\middle|\, f \in L^2(V) \text{ and } \sum_{x \in V} f(x) = 0 \right\} \\ &= \inf \left\{ \frac{\|df\|^2}{\|f\|^2} \,\middle|\, f \in L_0^2(V) \right\}.\end{aligned}$$

Proof. Since $L_0^2(X) = \{f \in L^2(X) | \sum_x f(x) = 0\}$ is orthogonal to the constants and Δ is self-adjoint, $\lambda_1(X)$ is the smallest eigenvalue of Δ on $L_0^2(X)$. So

$$\lambda_1(X) = \inf_{f \in L_0^2(X)} \left(\frac{\langle \Delta f, f \rangle}{\langle f, f \rangle} \right) = \inf_{f \in L_0^2(X)} \left(\frac{\langle df, df \rangle}{\langle f, f \rangle} \right). \qquad \square$$

The Cheeger constant of a graph was already defined in (1.1.3) to be:

$$h(X) = \inf_{A,B \subseteq V} \frac{|E(A,B)|}{\min(|A|, |B|)}$$

where the infimum runs over all the possibilities to make a disjoint partition $V = A \cup B$, and $E(A,B)$ is the set of edges connecting vertices in A to vertices in B.

In analogy to Cheeger's inequality (4.1.3), Dodziuk [Do] and Alon [A1] proved the following (see also Dodziuk-Kendall [DK]):

Proposition 4.2.4. *Let X be a finite graph with $d(x) \leq m$ for every vertex x. Then $\lambda_1(X) \geq \frac{h^2(X)}{2m}$.*

Proof. Let $g \in L_0^2(V)$ be an eigenfunction of Δ with respect to $\lambda = \lambda_1(X)$ and $\|g\| = 1$. So $\lambda = \langle dg, dg \rangle$. Let $V^+ = \{v \in V | g(v) > 0\}$ and

$$f(v) = \begin{cases} g(v) & v \in V^+ \\ 0 & \text{otherwise.} \end{cases}$$

Since we can replace g by $-g$, we may assume that $|V^+| \leq \frac{1}{2}|V|$. Now

$$\begin{aligned} \langle df, df \rangle &= \sum_{v \in V} f(v) \sum_{u \in V} \delta_{vu}(f(v) - f(u)) = \sum_{v \in V^+} g(v) \sum_{u \in V} \delta_{vu}(g(v) - f(u)) \\ &= \sum_{v \in V^+} g(v) \sum_{u \in V^+} \delta_{vu}(g(v) - g(u)) + \sum_{v \in V^+} \sum_{u \notin V^+} \delta_{vu} g(v) g(u). \end{aligned}$$

Since the last sum in the right-hand side is not positive, we get:

$$\begin{aligned} \langle df, df \rangle &\leq \sum_{v \in V^+} g(v) \sum_{u \in V} \delta_{vu}(g(v) - g(u)) = \sum_{v \in V^+} g(v) \Delta g(v) \\ &= \lambda_1(X) \sum_{v \in V^+} g^2(v) = \lambda_1(X) \langle f, f \rangle. \end{aligned} \qquad (*)$$

Consider the expression $A = \sum_{e \in E} |f^2(e^+) - f^2(e^-)|$. Clearly:

$$\begin{aligned} A &= \sum |f(x) + f(y)| \cdot |f(x) - f(y)| \\ &\leq (\sum |f(x) + f(y)|^2)^{\frac{1}{2}} \cdot (\sum |f(x) - f(y)|^2)^{\frac{1}{2}} \\ &\leq \sqrt{2} (\sum (f^2(x) + f^2(y)))^{\frac{1}{2}} \cdot \langle df, df \rangle^{\frac{1}{2}}. \end{aligned}$$

In $\sum(f^2(x) + f^2(y))$ every vertex contributes as many times as its degree. Hence

$$A \leq \sqrt{2m}\langle f, f\rangle^{1/2}\langle df, df\rangle^{1/2} \tag{$**$}$$

and by $(*)$, $A \leq \sqrt{2m\lambda_1(X)}\langle f, f\rangle$.

On the other hand, we can estimate A from below in terms of $\langle f, f\rangle$ as follows: Let $\beta_0 = 0 < \beta_1 < \cdots < \beta_r$ be the sequence of values of f union zero. Define $L_i = \{x \in V | f(x) \geq \beta_i\}$. Recall that since $|V^+| \leq \frac{1}{2}|V|$, for $i > 0$, $|L_i| \leq \frac{1}{2}|V|$. Now

$$A = \sum_{i=1}^{r} \sum_{f(x)=\beta_i} \sum_{f(y)<\beta_i} \delta_{xy}(f^2(x) - f^2(y)).$$

Note that $L_0 \supset L_1 \supset L_2 \supset \cdots \supset L_r$ and if $f(x) = \beta_i$ and x is connected by an edge to y with $f(y) = \beta_{i-j}$, then:

$$(f^2(x) - f^2(y)) = \beta_i^2 - \beta_{i-j}^2 = (\beta_i^2 - \beta_{i-1}^2) + (\beta_{i-1}^2 - \beta_{i-2}^2) + \cdots + (\beta_{i-j-1}^2 - \beta_{i-j}^2).$$

We can therefore write:

$$A = \sum_{i=1}^{r} \sum_{e \in \bar{\partial} L_i} \beta_i^2 - \beta_{i-1}^2 = \sum_{i=1}^{r} |\bar{\partial} L_i|(\beta_i^2 - \beta_{i-1}^2)$$

where $\bar{\partial} L_i$ is the set of all edges connecting vertices in L_i to vertices outside L_i, i.e., $\bar{\partial} L_i = E(L_i, X \backslash L_i)$ in the notation above. By the definition of Cheeger's constant $|\bar{\partial} L_i| \geq h(X) \cdot |L_i|$ for $i > 0$, and thus

$$\begin{aligned} A &\geq h(X) \sum_{i=1}^{r} |L_i|(\beta_i^2 - \beta_{i-1}^2) \\ &= h(X) \left(|L_r|\beta_r^2 + \sum_{i=1}^{r-1} \beta_i^2(|L_i| - |L_{i+1}|) \right). \end{aligned}$$

A vertex x is in $L_i \setminus L_{i+1}$ if and only if $f(x) = \beta_i$. Thus the above gives: $A \geq h(X)\langle g, g\rangle \geq h(X)\langle f, f\rangle$. Together with $(**)$ above, we deduce: $\lambda_1(X) \geq \frac{h^2(X)}{2m}$. □

The reader may notice that once again as in (3.15) and (3.3.7) the proof amounts to say that existence of «almost invariant» functions implies existence of «almost invariant» sets.

Similarly, in analogy to (4.1.4), Tanner [Ta] and Alon and Milman [AM] proved the following converse:

Proposition 4.2.5. $h(X) \geq \lambda_1(X)/2$.

Proof. Let $V = A \cup B$ be a disjoint union of two subsets of sizes a and b, respectively, $a + b = n = |V|$. Without loss of generality assume $a \le b$. Let

$$f(x) = \begin{cases} b & \text{if } x \in A \\ -a & \text{if } x \in B. \end{cases}$$

Then $f \in L_0^2(X)$, and hence

$$\lambda_1(X) \le \frac{\|df\|^2}{\|f\|^2} = \frac{n^2|E(A,B)|}{nab} \le 2 \cdot \frac{|E(A,B)|}{a}.$$

This is true for every such partition which shows that $h(X) \ge \frac{1}{2}\lambda_1(X)$. □

Propositions (1.1.4), (4.2.4) and (4.2.5) indicate that the problem of finding expanders is essentially equivalent to bound λ_1 from below. The following result shows that there is a limitation to the possibility of bounding λ_1 for an infinite family of graphs.

Proposition 4.2.6. (Alon-Boppana) *Let $X_{n,k}$ be a family of k-regular graphs where k is fixed and n is the number of vertices of $X_{n,k}$ is going to infinity. Then:*

$$\lim_{n \to \infty} \sup \lambda_1(X_{n,k}) \le k - 2\sqrt{k-1}.$$

We will not prove this proposition here. Rather we will come back to (a slightly weaker version of) it in (4.5.5). Here we mention some extensions of this result:

Theorem 4.2.7. (Greenberg [Gg]) *Let X be an infinite connected graph and $\mathcal{L}$ the family of finite graphs covered by X. (Where here we mean that $Y \in \mathcal{L}$ if Y is finite and there exists a group Γ_Y on X with $\Gamma_Y \setminus X = Y$.) Let φ be the spectral radius of the averaging operator $\delta = \delta_X$ acting on $L^2(X)$. Then for every $\epsilon > 0$ there exists $c = c(X, \epsilon)$, $0 < c < 1$, such that if $Y \in \mathcal{L}$ with $|Y| = n$, then at least cn of the eigenvalues λ of δ_Y satisfy $\lambda \ge \varphi - \epsilon$.*

The theorem generalizes an unpublished result of Serre concerning the case $X = X_k$: the k-regular tree.

Corollary 4.2.8. *With the notations of (4.2.7), fix an integer $r \ge 0$. Then*

$$\lim_{|Y| \to \infty} \inf \mu_r(Y) \ge \varphi$$

when $\mu_0(Y) > \mu_1(Y) \ge \mu_2(Y) \ge \cdots \ge \mu_{n-1}(Y)$ denote the eigenvalues of δ_Y – the adjacency matrix of Y.

The last corollary is due to Burger [Bur4]. It is an analogue to a result of Cheng [Che] in the geometric context. As the spectral radius of δ_{X_k} is $2\sqrt{k-1}$, (4.2.8) is a far-reaching generalization of (4.2.6).

4.3 Eigenvalues, isoperimetric inequalities and representations

Keeping in mind (1.1.4), the results of the previous section, (4.2.4) and (4.2.5), we show that the question of finding families of k-regular expanders is essentially equivalent to presenting a family of k-regular graphs X_i for which $\lambda_1(X_i)$ is bounded away from 0. We now show that, moreover, if our graphs are Cayley graphs of quotients of a fixed finitely generated group, this is also equivalent to a «relative property T». (We remark that a family of k-regular graphs, for k fixed and even, can always be presented as a family of Schreier graphs of quotients of some infinite group. This can be deduced from [AG]. So the situation we study in this section is quite general.)

Before making the statement precise, let's define:

Definition 4.3.1. *Let Γ be a finitely generated group and $\mathcal{L} = \{N_i\}$ a family of finite index normal subgroups of Γ. Let $R = \{\varphi \in \hat{\Gamma} | \ker\varphi$ contains N_i for some $i\}$. We say that Γ has property (τ) with respect to the family $\mathcal{L}$ if the trivial representation is isolated in the set R (i.e., Γ has property $(T : R)$ in the terminology of (3.1.10)). We say that Γ has property (τ) if it has this property with respect to the family of all finite index normal subgroups.*

Theorem 4.3.2. *Let Γ be a finitely generated group generated by a finite symmetric set of generators S. Let $L = \{N_i\}$ be a family of finite index normal subgroups of Γ. Then the following conditions are equivalent:*

i. *Γ has property (τ) with respect to L (i.e., there exists an $\varepsilon_1 > 0$ such that if (H, ρ) is a non-trivial unitary irreducible representation of Γ whose kernel contains N_i for some i, then for every $v \in H$ with $\|v\| = 1$, there exists $s \in S$ such that $\|\rho(s)v - v\| \geq \varepsilon_1$).*
ii. *There exists $\varepsilon_2 > 0$ such that all the Cayley graphs $X_i = X(\Gamma/N_i, S)$ are $([\Gamma : N_i], |S|, \varepsilon_2)$-expanders.*
iii. *There exists $\varepsilon_3 > 0$ such that $h(X_i) \geq \varepsilon_3$.*
iv. *There exists $\varepsilon_4 > 0$ such that $\lambda_1(X_i) \geq \varepsilon_4$.*

   If in addition $\Gamma = \pi_1(M)$ for some compact Riemannian manifold M, and M_i are the finite sheeted coverings corresponding to the finite index subgroups N_i, then the above conditions are also equivalent to each of the following:

v. *There exists $\varepsilon_5 > 0$ such that $h(M_i) \geq \varepsilon_5$.*
vi. *There exists $\varepsilon_6 > 0$ such that $\lambda_1(M_i) \geq \varepsilon_6$.*

Proof. (i) $\Rightarrow$ (ii) is just like the proof that property (T) gives expanders in proposition (3.3.1). (See also Remark (3.3.4).)

(ii) $\Leftrightarrow$ (iii) is contained in (1.1.4).

(iii) $\Leftrightarrow$ (iv) is Doziuk-Alon's result (4.2.4) and Tanner-Alon-Milman (4.2.5).

To prove (iv) $\Rightarrow$ (i): Recall that if ρ factors through the finite quotient Γ/N_i, then ρ appears in the right regular representation of Γ on the group algebra $\mathbb{C}[\Gamma/N_i] \simeq L^2(\Gamma/N_i)$, and indeed ρ appears in $L_0^2(X_i)$ since ρ is nontrivial. Moreover, if A is the element of the group algebra of Γ (or Γ/N_i) given by $\sum_{s\in S} s$, then Δ acts on $L^2(X_i) = \mathbb{C}[\Gamma/N_i]$ as a multiplication from the right by $k \cdot e - A$ where e is the identity element of Γ and $k = |S|$. By (iv), we have that for every $f \in L_0^2(X_i)$ with $\|f\| = 1$, $\|\Delta f\| \geq \varepsilon_4$. Thinking of f as an element of $\mathbb{C}[\Gamma/N_i]$, we have:

$$\varepsilon_4 \leq \|\Delta f\| = \|f \cdot \sum_{s\in S}(e-s)\| = \|\sum_s f\cdot(e-s)\| \leq \sum_s \|f\cdot(e-s)\|.$$

This implies that at least for one $s \in S$, $\frac{\varepsilon_4}{k} \leq \|f\cdot(e-s)\| = \|f - f\cdot s\| = \|f - R(s)f\|$. Here $R(s)$ is the right regular representation of Γ on $L_0^2(\Gamma/N_i)$. So $\varepsilon_1 = \frac{\varepsilon_4}{k}$ works for $R(s)$ and in particular for ρ.

The equivalence of (v) and (vi) follows from (4.1.3) and (4.1.4).

To prove that (v) implies (iii), we observe first that condition (iii) (as well as all other conditions) are independent of the choice of generators (yet the values of the $\varepsilon - s$ do depend on S), so it suffices to prove (iii) for some set of generators.

Let U be the universal covering space of M. Then Γ acts on U and $M = U/\Gamma$. Fix a compact closed fundamental domain $\mathcal{F}$ for Γ with «faces» $C_1, \dots, C_r$. Then take $\gamma_j \in \Gamma$ such that $\gamma_j(\mathcal{F}) \cap \mathcal{F} = C_j$. It is well known that $S = \{\gamma_j\}$ generates Γ. Now every one of the finite sheeted coverings M_i of M is also covered by U with $\varphi_i : U \mapsto M_i$ the covering map. Then $\overline{\mathcal{F}}_i = \varphi_i(\mathcal{F})$ is a fundamental domain for the action of Γ/N_i on M_i, and $M_i/(\Gamma/N_i) = M$. We claim that the Cayley graph $X_i = X(\Gamma/N_i, S)$ can be «drawn» in a natural way on M_i as follows: The Γ/N_i-translations of $\overline{\mathcal{F}}_i$ will be the vertices of the graph, and two vertices are adjacent if and only if they have a common face. Now, any partition of the graph into two disjoint sets A and B induces a partition of the manifold into two subsets whose intersection (boundary) is the union of the faces corresponding to the edges $E(A, B)$ – going from elements of A to elements of B. We can now conclude (since the volume of $\mathcal{F}$ is fixed and the areas of the faces are bounded for all i) that a lower bound on $h(M_i)$ gives rise to a lower bound on $h(X(\Gamma/N_i), S)$. (See also (4.4.3) below for a concrete computation.)

The opposite direction ((iii) implies (v)) is more complicated; here one should show how an arbitrary hypersurface dividing the manifold M_i implies the same kind of division for the graph X_i. For the proof of this, the reader is referred to Brooks [Br2] (see also [Br1]). □

Examples 4.3.3.

A. Let Γ be a finitely generated Kazhdan group; then it has property (τ) and hence all the properties of the theorem with respect to all the finite index normal subgroups.

B. If Γ is a finitely generated, residually finite (i.e., the intersection of the normal subgroups of finite index is trivial) amenable group then Γ has (τ) if and only if Γ is finite. This follows from (3.3.7) and (4.3.2).

C. Let F be the free group on two generators x and y, $F = F(x,y)$. For every integer m, let φ_m be the homomorphism from F onto the symmetric group S_m, sending x to $\gamma = (1,2)$ and y to $\sigma = (1,2,3,\dots,m)$. Let N_m be the kernel of φ_m. We claim that F does not have property (τ) with respect to the family $\{N_m\}$. To see this, let ρ_m be the m-dimensional representation of S_m (and hence of F) defined by:
For $v = (v_1,\dots,v_m) \in V = \mathbb{C}^m$ and $\beta \in S_m$,

$$\rho_m(\beta)(v) = (v_{\beta^{-1}(1)},\dots,v_{\beta^{-1}(m)}).$$

It is well known that the restriction of ρ_m to $V_0 = \{v = (v_1,\dots,v_m) | \sum v_i = 0\}$ is an irreducible representation. Let $\eta = e^{\frac{2\pi i}{m}}$. Then $v = (1,\eta,\dots,\eta^{m-1})$ is in V_0 and $\|v\| = \sqrt{m}$. Now:

$$\|\gamma v - v\| = \|(\eta - 1, 1 - \eta, 0,\dots,0)\| = \sqrt{2}|1-\eta| \xrightarrow[m\to\infty]{} 0$$

and

$$\|\sigma v - v\| = \|(\eta^{m-1} - 1, 1 - \eta, \eta - \eta^2,\dots,\eta^{m-2} - \eta^{m-1})\| = \sqrt{m}|1-\eta|.$$

So, $\lim_{m\to\infty} \frac{\|\sigma v - v\|}{\|v\|} = 0$. This proves that F does not have property (τ) with respect to $\{N_m\}$.
As a corollary we deduce that $X_m = X(S_m, \{\gamma,\sigma\})$ – the Cayley graphs of S_m with respect to the generators τ and σ are not expanders. This can be proved directly by looking at the subset A_m of X_m defined by:

$$A_m = \{\beta \in S_m | 1 \le \beta(1) \le \frac{m}{2}\}.$$

Then $\tau \cdot A_m = A_m$ and $\frac{|\sigma \cdot A_m \Delta A_m|}{|A_m|} \simeq \frac{1}{m} \xrightarrow[m\to\infty]{} > 0$, which shows that X_m is not a family of expanders.
In fact, in (8.1.6), we shall show that the diameter of X_m is $O(m^2)$ while it is easy to see that expanders should have logarithmic diameter (see also (7.3.11)), which in case of S_m should be $O(\log m!) = O(m \log m)$.
It is a very interesting open problem to find (if possible) a set of k generators (k fixed) for every S_m which will make the symmetric groups into a family of expanders. (See also Problem (3.3.3) and Section 8.2.)

D. Probably the most interesting example of a group Γ with a family of normal subgroups satisfying (τ) is $\Gamma = SL_2(\mathbb{Z})$ and $N_m = \Gamma(m) = \ker(SL_2(\mathbb{Z})$

$\mapsto SL_2(\mathbb{Z}/m\mathbb{Z})$. This follows from Selberg's Theorem, which we describe in the next section. Here we only mention that Selberg's Theorem has various extensions by Gelbart-Jacquet [GJ], Sarnak [Sa1], Elstrodt-Gruenwald-Mennicke [EGM], Li-Piatetski-Shapiro-Sarnak [LPSS] and Li [Li]. All these papers together imply that if Γ is an arithmetic group in a simple real Lie group, then Γ has property (τ) with respect to the family of congruence subgroups.

E. Another relevant application of Selberg's Theorem is that the group $\Gamma = SL_2(\mathbb{Z}\left[\frac{1}{p}\right])$ has property (τ) but not property (T). The second statement is clear since Γ is dense in $SL_2(\mathbb{R})$. To prove the first: Let $\Gamma_1 = SL_2(\mathbb{Z}) \hookrightarrow \Gamma$ and S_1 a set of generators for Γ_1. Extend it to a set of generators for Γ. Now, by the affirmative solution to the congruence subgroup problem for Γ (see [S2]), every finite quotient of Γ factors through a congruence subgroup mod m (for m prime to p). But then the image of Γ_1 in $SL_2(\mathbb{Z}\left[\frac{1}{p}\right]/m\mathbb{Z}\left[\frac{1}{p}\right]) = SL_2(\mathbb{Z}/m\mathbb{Z})$ is onto. Thus by Selberg's Theorem every finite quotient of Γ is an expander with respect to S_1 and therefore also with respect to S. By theorem (4.3.2), Γ has property (τ).

The results of Gelbart-Jacquet [GJ] and Sarnak [Sa1] give other such examples, e.g., if $\mathcal{O}$ is the ring of integers in a quadratic real number field (e.g., $\mathcal{O} = \mathbb{Z}[\sqrt{2}]$) then $SL_2(\mathcal{O})$ has property (τ) but not (T).

In Lubotzky-Zimmer [LZ] it is proved that if Γ is an irreducible lattice in a non-trivial product of two non-compact locally compact groups $G_1 \times G_2$ where G_1 is Kazhdan (but G_2 is not), then Γ has property (τ) (but not (T)). This can be used to give many examples of groups with property (τ) and not (T) without appealing to deep number theory as the previous examples do.

4.4 Selberg Theorem $\lambda_1 \geq \frac{3}{16}$ and expanders

A case for which Theorem (4.3.2) can be applied is the group $\Gamma = SL_2(\mathbb{Z})$ due to a theorem of Selberg [Sl]. This will enable us to get a few additional families of interesting expanders.

Let $\Gamma = SL_2(\mathbb{Z})$. As a subgroup of $G = SL_2(\mathbb{R})$, it acts on the upper half-plane $\mathbb{H} = \{z = x+iy \in |x, y \in \mathbb{R}, y > 0\}$ by: $g(z) = \frac{az+b}{cz+d}$ when $g = \begin{pmatrix} a & b \\ c & d \end{pmatrix} \in G$. The action of G preserves the hyperbolic metric of $\mathbb{H}$, $ds^2 = \frac{1}{y^2}(dx^2 + dy^2)$, and it, therefore, commutes with its Laplacian $\Delta = -y^2\left(\frac{\partial^2}{\partial x^2} + \frac{\partial^2}{\partial y^2}\right)$. Thus, if Γ' is any discrete subgroup of G, then Δ is well defined on the quotient $M = \Gamma' \setminus \mathbb{H}$

and it is the Laplacian of M. This is in particular true for subgroups of finite index of Γ. Such a subgroup is called a congruence subgroup if it contains

$$\begin{aligned}\Gamma(m) &= \ker(SL_2(\mathbb{Z}) \mapsto SL_2(\mathbb{Z}/m\mathbb{Z})) \\ &= \left\{ \begin{pmatrix} a & b \\ c & d \end{pmatrix} \in \Gamma \,\middle|\, \begin{matrix} a \equiv d \equiv 1 (\bmod m) \\ b \equiv c \equiv 0 (\bmod m) \end{matrix} \right\}\end{aligned}$$

for some $m \neq 0$.

Theorem 4.4.1. (Selberg [Se]) *If Γ' is a congruence subgroup of $\Gamma = SL_2(\mathbb{Z})$, then $\lambda_1(\Gamma' \setminus \mathbb{H}) \geq \frac{3}{16}$.*

Selberg conjectured that in fact $\lambda_1(\Gamma' \setminus \mathbb{H}) \geq \frac{1}{4}$. We will refer to it as «Selberg conjecture».

Theorem 4.4.2.

i. *Let p be a prime number and let $\tau = \begin{pmatrix} 0 & 1 \\ -1 & 0 \end{pmatrix}$ and $\psi = \begin{pmatrix} 1 & 1 \\ 0 & 1 \end{pmatrix}$ be the 2×2 matrices in $SL_2(p)$. Then the Cayley graphs $X_p = X(PSL_2(p); \{\tau, \psi\})$, when p runs over all primes, form a family of expanders.*
ii. *Let $Y_p = \{0, 1, 2, \ldots, p-1, \infty\}$. Connect every $z \in Y_p$ to $z \pm 1$ and to $-\frac{1}{z}$, then the resulting cubic graphs form a family of expanders.*

Proof. i. This part follows from the implication (vi)$\Rightarrow$(ii) in Theorem (4.3.2) and Theorem (4.4.1) (see Remark 5 in Section (4.6)).

ii. This part follows from part (i) since Y_p is a quotient of X_p by the action of the subgroup $\{\begin{pmatrix} a & b \\ o & c \end{pmatrix}\}$ of $SL_2(p)$. □

The method of proof of Theorem (4.4.2) is in fact effective, and one can give a lower bound for $\lambda_1(X_p)$ as for the expansion coefficients. In an early version of [LPS1] it was shown, for example, that $\lambda_1(Y_p) \geq 0.003$. Here we will present some different examples which will illustrate how the transition from the geometric Cheeger constant of the manifold to the combinatorial Cheeger constant of the graphs (i.e., the implication (v)$\Rightarrow$(iii) in Theorem (4.3.2)) can be done in an effective way. (The same remark applies to all the other implications of Theorem (4.3.2) and in particular to (vi)$\Rightarrow$(iv)).

Before doing this we mention that due to Jacquet-Langlands correspondence [JL], the Selberg result is automatically valid for congruence subgroups of cocompact arithmetic lattices in $SL_2(\mathbb{R})$. (We will explain this point later in Chapter 6; see Theorem (6.3.4).) Let's take such a cocompact lattice, for example, the triangle group $T = (4, 4, 4)$. More precisely, let D be a hyperbolic triangle in $\mathbb{H}$ of angles $\alpha = \beta = \gamma = \frac{\pi}{4}$ and edges a, b and c, respectively (i.e., a is opposite α, etc.). Let A, B and C be the reflections with respect to a, b and c, respectively. $\widetilde{T}$ will be the group generated by A, B and C and T the index two subgroup generated by AB, BC and CA. The last three are in $PSL_2(\mathbb{R})$, and so T is a subgroup of $PSL_2(\mathbb{R})$. It is well known that T is a cocompact lattice (see [Be, Section 10.6]).

By the results of Takeuchi [T], for the specific values $\alpha = \beta = \gamma = \frac{\pi}{4}$, and the group T is arithmetic.

As mentioned above, if N is a congruence subgroup of T, then $\lambda_1(N \setminus \mathbb{H}) \geq \frac{3}{16}$. Let $M = M_N$ be the manifold $N \setminus \mathbb{H}$. Then M comes with a natural triangulation, by triangles of type $\left(\frac{\pi}{4}, \frac{\pi}{4}, \frac{\pi}{4}\right)$. Define the dual graph of it to be the graph whose vertices are the triangles and where two vertices are adjacent if the corresponding triangles have a common face. Denote this graph by $X(M_N)$.

Theorem 4.4.3. *When N runs over the congruence subgroups of T, then* $h(X(M_N)) \geq 0.035$. *In particular these are* $([\widetilde{T} : N], 3, 0.01)$*-expanders.*

Proof. We first note that the graph $X(M_N)$ is just the Cayley graph of $\widetilde{T}/N$ with respect to the three reflections as generators. So Theorem (4.3.2) and the fact that $\lambda_1(M_N) \geq \frac{3}{16}$ imply that these are expanders. To compute the constants involved: Recall that Buser's Theorem (4.1.4) implies that $h(M) \geq \frac{1}{10}(\sqrt{23/8} - 1) \approx 0.0695$.

Now the area of the basic triangle $\left(\frac{\pi}{4}, \frac{\pi}{4}, \frac{\pi}{4}\right)$ is $V = \pi - \left(\frac{\pi}{4} + \frac{\pi}{4} + \frac{\pi}{4}\right) = \frac{\pi}{4}$ (see [Be, Theorem 7.13.1]). The second Cosine Rule of hyperbolic geometry yields $\cosh(c) = \frac{\cos\alpha\cos\beta + \cos\gamma}{\sin\alpha\sin\beta}$ (see [Be, Section 7.12]). For our case this says that $\cosh(c) = 1 + \sqrt{2}$ and so $a = b = c \approx 1.52$. Now let A and B be disjoint sets of vertices of the graph which form a partition of it. Let $\overline{A}$ and $\overline{B}$ be the subsets of M such that $\overline{A}$ is the union of the triangles corresponding to the vertices in A (and similarly with $\overline{B}$). The «hypersurface» separating $\overline{A}$ and $\overline{B}$ consists of the lines corresponding to the set of edges $E(A, B)$ going from A to B on the graph. The length of this hypersurface is therefore $1.52|E(A, B)|$, and the volume of A is $\frac{\pi}{4}|A|$. Since $h(M) \geq 0.0695$, we deduce that $\frac{1.52|E(A,B)|}{\frac{\pi}{4}|A|} \geq 0.0695$. This implies that $\frac{|E(A,B)|}{|A|} \geq 0.035$ and the theorem is proved. □

One can go ahead and present the graphs in the last theorem in a more explicit way by computing the generators of $\widetilde{T}/N$. But, we will not go to this trouble since we eventually will have much better expanders. The point of this section (and in particular Theorems (4.4.2) and (4.4.3)) is, however, to illustrate that unlike property (T), which gives expanders but without an estimation of the expansion coefficient, Selberg's Theorem comes with a number! This number is not so good since we lose when translating the bound on the geometric Laplacian to be a bound on the approximated graphs. This is the motivation to look for a situation similar to Selberg's, but where the «manifolds» are already graphs. This will be done in chapters 6 and 7. Before that, in Chapter 5, we shall try to understand Selberg Theorem from a representation theoretic point of view and to understand how it stands with respect to property (T).

4.5 Random walks on k-regular graphs; Ramanujan graphs

Let X be a k-regular graph (finite or infinite). Consider the random walk on X in which every step consists of moving with probability $\frac{1}{k}$ along one of the edges coming out of the vertex. This random walk defines a Markov chain whose possible states are the vertices of X. The Markov chain defines an operator $M : L^2(X) \mapsto L^2(X)$ defined by $(Mf)(x) = \frac{1}{k}\sum_{y\sim x} f(y)$ where the sum runs over the neighbors of x (with multiplicity if needed). There is a clear connection between M and the Laplacian Δ of X. In this section we survey some of the known results on this random walk and the eigenvalues of M and relate them to what interests us: the eigenvalues of Δ.

An important special case is a Cayley graph X of a finitely generated group. The theory here was initiated by Kesten [Ke1]. For proofs the reader is referred to that paper as well as to the paper of Buck [Bc], on which our exposition here is based. So M is defined as above. The spectrum of M, spec(M), is the set of all complex numbers λ such that $M - \lambda I$ does not have an inverse with finite norm. (A norm of an operator T is as usual $\|T\| = \sup\{\|Ty\| \mid y \in L^2(X) \text{ and } \|y\| = 1\}$). The spectral radius of M is defined as $\rho(M) = \sup\{|\lambda| \mid \lambda \in \text{spec}(M)\}$ and it is equal to the norm of M. Clearly $\|M\| \leq 1$.

Proposition 4.5.1. *Fix $x_0 \in X$ (call it the origin). Let r_n be the probability of the random walk being at the origin at time n having started there at time 0. Then,*

$$\|M\| = \limsup (r_n)^{1/n}.$$

$\|M\|$ is also equal to the reciprocal of the radius of convergence of the return generating function $R(z) = \sum_{r=0}^{\infty} r_n z^n$.

Proposition 4.5.2. *Let $X = X_k$ be the homogeneous k-regular tree, and M as above; then $\|M\| = \frac{2\sqrt{k-1}}{k}$.*

Proof. Fix the vertex v_0 in X and a vertex y_0 adjacent to x_0. Let $Q(z)$ be the first return generating function, i.e., it is the power series whose z^n coefficient, for $n > 0$, is the probability that the random walk wanders back to the origin x_0 for the first time at time n. Let $T(z)$ be the generating function giving the probability of first reaching the origin x_0 after a number of steps from the starting point y_0. Thus we have $Q(z) = \sum_{y\sim x_0} \frac{1}{k} \cdot zT(z) = zT(z)$ and $R(z) = \frac{1}{1-Q(z)}$ (since $R(z) = 1 + Q(z)R(z)$).

Starting at a point y_1, instead of y_0, at distance m from x_0, the generating function is obtained by raising T to the power m. From the point y_0, there are $k-1$ ways to go to a point twice removed from x_0, and only one way to go directly to x_0. Thus T satisfies the identity $T(z) = \frac{z}{k} + \frac{k-1}{k} zT^2(z)$. Solving for T we obtain: $T(z) = \frac{k \pm \sqrt{k^2 - 4(k-1)z^2}}{2(k-1)z}$. Notice that we can think of the value of

$T(z)$, for $0 \leq z \leq 1$, as the probability of ever reaching x_0, starting at y_0, given that we follow the random walk but also have the probability $1 - z$ of dying at each step. Now, since $T(z) \leq 1$ for $0 \leq z \leq 1$, we must take only the minus sign above, as the other root will always be at least 1 when $k \geq 2$. The numerator in the above expression has a zero of order 2 at the origin, considered as a function of a complex variable, so the radius of convergence of T is the distance to the branch point of the square root function. The branch point occurs at the zero of the discriminant, namely $z_0 = k/2\sqrt{k-1}$. Thus by the previous proposition and since $R(z) = \frac{1}{1-zT(z)}$, we deduce $\|M\| = \frac{2\sqrt{k-1}}{k}$. □

For example, if F is a free group on free r generators, its Cayley graph is a $2r$-regular tree and the norm is therefore $\frac{2\sqrt{2r-1}}{2r}$. Kesten ([Ke1], [Ke2]) proved:

Theorem 4.5.3. *Let Γ be a group and $S = \{a_1, \ldots, a_r, b_1, \ldots, b_t\}$ a finite set of generators of Γ such that $b_1, \ldots, b_t$ are of order 2 while $a_1, \ldots, a_r$ are of order greater than 2. Let $X = X(\Gamma; S)$ be the Cayley graph of Γ with respect to S and M the operator of the random walk on X (i.e., $M = \frac{1}{2r+t}\delta_X$). Then:*

1. *Γ is amenable if and only if $\|M\| = 1$.*
2. *Γ is a free product of a free group on the r generators $a_1, \ldots, a_r$ and the groups of order 2, (i.e., $\Gamma = \langle a_1 \rangle * \langle a_2 \rangle * \cdots * \langle a_r \rangle * \langle b_1 \rangle * \cdots * \langle b_t \rangle$) if and only if $\|M\| = \frac{2\sqrt{2r+t-1}}{2r+t}$. (This happens if and only if X is a $(2r+t)$-regular tree).*

Proposition (4.5.2) has applications also to the study of M for finite k-regular graphs. The following proposition was announced by Alon-Boppana (cf. [A1]). Proofs are given in [Bc] or [LPS2].

Proposition 4.5.4. *Let $X = X_{n,k}$ be a family of k-regular graphs on n vertices, with n is going to infinity and fixed k. When X is a bipartite graph, write $X = I \cup O$ where I is the input set and O is the output set; if X is not bipartite, write $X = I = O$.*

Now, let

$$L_0^2(X_{n,k}) = \left\{ f \in L^2(X_{n,k}) \,\middle|\, \sum_{x \in O} f(x) = \sum_{x \in I} f(x) = 0 \right\}.$$

Let $N = M_n^o$ be the Markov chain operator restricted to $L_0^2(X_{n,k})$. Then

$$\liminf_{n \to \infty} \|M_n^o\| \geq \frac{2\sqrt{k-1}}{k}.$$

Proof. Let $\widetilde{X}$ be the infinite k-regular tree and $\widetilde{M}$ its random walk operator. If diameter$(X) \geq 2r+1$, choose vertices x_1 and x_2 in I of distance $\geq 2r+1$. Let $B_r(x_i)$ be the ball of radius r around x_i, $i=1,2$. Then $B_r(x_1) \cap B_r(x_2) = \emptyset$. Define the functions $\delta_i = \chi_{\{x_i\}}$ and $f = \delta_1 - \delta_2$. Then $f \in L_0^2(V)$. If e is the origin of $\widetilde{X}$, write $\delta_e = \chi_{\{e\}}$. Now, if $\widetilde{X}$ covers X and $\widetilde{X}$ is homogeneous, it follows that for every r and for $i=1,2$

$$\|\widetilde{M}^r(\delta_e)\| \leq \|M^r(\delta_i)\|.$$

Moreover, for r as above, $\|M^r(f)\|^2 = \|M^r(\delta_1)\|^2 + \|M^r(\delta_2)\|^2$ since $\|M^r(\delta_i)\|$ is supported on $B_r(x_i)$ and $B_r(x_1) \cap B_r(x_2) = \emptyset$. Putting this together:

$$\begin{aligned} 2\|N\|^{2r} \geq \|N\|^{2r} \cdot \|f\|^2 \geq \|N^r(f)\|^2 &= \|M^r(f)\|^2 \\ &= \|M^r(\delta_1)\|^2 + \|M^r(\delta_2)\|^2 \geq \|\widetilde{M}^r(\delta_e)\|^2. \end{aligned}$$

Thus $\|N\| \geq \|\widetilde{M}^r(\delta_e)\|^{1/r} 2^{-1/2r}$.

But we remember that the right-hand side of the last inequality converges to $\|\widetilde{M}\|$ when $r \to \infty$. We also note that when $n \to \infty$ the diameter also goes to infinity (k is fixed!) and hence $\liminf \|M_n^o\| \geq \|\widetilde{M}\| = \frac{2\sqrt{k-1}}{k}$. □

Now, for a k-regular graph, M is nothing more than $\frac{1}{k}\delta$ where δ is its adjacency matrix.

Corollary 4.5.5. (Alon-Boppana) *Let $X_{n,k}$ be a family of k-regular connected graphs. Let*

$$\lambda(X_{n,k}) = \max\{|\lambda| \mid \lambda \text{ is an eigenvalue of } \delta_{X_{n,k}},\ \lambda \neq \pm k\}.$$

Then $\liminf_{n\to\infty} \lambda(X_{n,k}) \geq 2\sqrt{k-1}$.

For reasons to be clearer later, we make the following definition, following [LPS1, LPS2]:

Definition 4.5.6. *A k-regular graph $X = X_{n,k}$ is called a Ramanujan graph if for every eigenvalue λ of the adjacency matrix δ_X either $\lambda = \pm k$ or $|\lambda| \leq 2\sqrt{k-1}$.*

Recall that for a k-regular graph X, k is always an eigenvalue. It is the largest eigenvalue and of multiplicity one if X is connected. On the other hand, $-k$ is an eigenvalue if and only if X is bi-partite. Thus Ramanujan graphs are those k-regular graphs which satisfy the strongest asymptotic bound on their eigenvalues. Much of the rest of the notes will be devoted to an explicit construction of such graphs. We shall also put them in a more general perspective, explaining the connection between them and Ramanujan conjecture.

In Proposition (3.3.1) we constructed expanders as quotient graphs of a group Γ with property (T). The next proposition shows that those expanders are not Ramanujan, except for possibly finitely many of them:

Proposition 4.5.7. *Let Γ be a group generated by a set S of ℓ generators $x_1, \dots, x_\ell$ such that the first s of them $x_1, \dots, x_s$ (possibly $s = 0$) are of order 2. (So, the Cayley graph $X(\Gamma; S)$ is a $k = 2\ell - s$ regular graph.)*

Assume there exists infinitely many finite index normal subgroups N_i of Γ such that $N_i \cap S = \emptyset$ and $X(\Gamma/N_i; S)$ are k-regular Ramanujan graphs. Then $X(\Gamma; S)$ is the k-regular tree, and hence Γ is a free product of s cyclic groups of order 2 and a free group on $\ell - s$ generators. In particular, Γ contains a finite index free group and hence it does not have property (T).

Proof. $X = X(\Gamma; S)$ is a covering graph for all the finite Cayley graphs $X(\Gamma/N_i; S)$. It follows from Corollary (4.2.8) that the norm of the random walk on X is $2\sqrt{k-1}$. Thus, by Theorem (4.5.3), X is a tree. This means that there is no closed path in X, i.e., any reduced word in S is non-trivial in Γ. This is equivalent to say that Γ is the free product of the cyclic groups $\Gamma = \langle x_1 \rangle * \langle x_2 \rangle * \cdots * \langle x_\ell \rangle$ where x_i is either of order 2 or of infinite order. The kernel of the homomorphism $\phi\colon \Gamma \to \mathbb{Z}/2\mathbb{Z}$ with $\phi(x_i) = 1$ for $i = 1, \dots, l$ is a free subgroup of finite index (by the Kurosh subgroup theorem). This completes all parts of the proposition. □

We will indeed construct Ramanujan graphs as quotients of a free group, but of course not all quotients of a free group are Ramanujan. They are even not expanders, as can be seen by looking at these quotients factoring through the commutator quotient, or more generally through any given amenable group (by Proposition (3.3.7)).

We end up this section with an exotic property of Ramanujan graphs, which also indicate the naturality of this definition.

Let X be a finite k-regular graph, with $k = q + 1$. For every homotopy class C of closed paths, choose one of a minimal length denoted $\ell(C)$. A class C is primitive if it is not a proper power of another class in the fundamental group of X. Let

$$Z_X(s) = \prod_{\text{primitive classes } C} (1 - q^{-s \cdot \ell(C)})^{-1}.$$

$Z_X(s)$ is called the «zeta function» of the graph X. The following proposition is due to Ihara [Ih1] (see also Sunada [Sd] and Hashimoto [Ha]).

Proposition 4.5.8. *If X is a k-regular graph, $k = q + 1$. Then*

$$Z_X(s) = (1 - u^2)^{-(r-1)} \det(I - \delta u + q u^2)^{-1}$$

where $u = q^{-s}$, $r = \operatorname{rank} H^1(X, \mathbb{Z})$ and δ is the adjacency matrix of X.

Corollary 4.5.9. *Let X be a k-regular graph. Then X is a Ramanujan graph if and only if $Z_X(s)$ satisfies «the Riemann hypotheses», i.e., all the poles of $Z_X(s)$ in $0 < \operatorname{Re}(s) < 1$ lie on the line $\operatorname{Re}(s) = \frac{1}{2}$.*

Proof. Let $\phi(z)$ be the characteristic polynomial of δ_X, i.e., $\phi(z) = \det(zI - \delta_X)$. Then by (4.5.8), $Z_X(s)^{-1} = (1-u^2)^{r-1}u^n\phi\left(\frac{1+qu^2}{u}\right)$, where n is the number of vertices of X. Let $z = \frac{1+qu^2}{u}$ then $qu^2 - zu + 1 = 0$ and $u = \frac{z\pm\sqrt{z^2-4q}}{2q}$. If $z = \pm(q+1)$ then $u = \pm 1, \pm\frac{1}{q}$ and $\mathrm{Re}(s) = 0$ or $\mathrm{Re}(s) = 1$. «X is Ramanujan» means that for any other z (s.t. $\phi(z) = 0$), $|z| \leq 2\sqrt{q}$ so $u = \pm q^{-\frac{1}{2}}$ (and $\mathrm{Re}(s) = \frac{1}{2}$) or u is nonreal. Since δ_X is symmetric, z is real, hence $\frac{z\bar{u}}{\bar{u}} = \frac{(1+qu^2)\bar{u}}{u\bar{u}} = \frac{\bar{u}+q|u|^2u}{u^2}$ is also real, and thus $q|u|^2 = 1$, i.e., $|u| = q^{-1/2}$, i.e., $\mathrm{Re}(s) = \frac{1}{2}$. □

For more in this direction, see Sunada [Sd].

4.6 Notes

1. The Laplacian operator on manifolds with various boundary conditions has received a lot of attention for its importance in various problems in physics, differential geometry and representation theory. In particular, many authors discussed the influence of bounds on its eigenvalues on the geometry of the manifolds and vice versa. The results of Cheeger and Buser mentioned in Section (4.1) are just two examples. More on this can be found in [Ber], [Ch] and [BGM]. It seems to be of interest to transfer more results of this type to graphs (cf. [Br4]). One can also attempt to interpret some graph theoretic properties for manifolds. A challenging example: the chromatic number of a graph is a graph theoretic invariant which is related to eigenvalues of the Laplacian (see [Ho]). Define the «chromatic number» of a manifold and relate it to its eigenvalues.
2. Alon [A1] has a different version of (4.2.4) which gives a lower bound for $\lambda_1(X)$ by means of the expansion of the graph which is independent of the degree. We preferred to give here Dodziuk's version since its proof follows closely Cheeger's proof for manifolds. The proof also illustrates how to prove that if every set has large boundary (or is moved «a lot» by the generators – in case this is a Cayley graph) then every function is moved «a lot» by the Laplacian. It is interesting to compare this with the proof of (3.1.5). An improvement of Dodziuk's version is given by Dodziuk and Kendall [KD], but this also uses Cheeger constant $h(X)$ and not $c(X)$ – the expansion constant. Various improvements of (4.2.4) can be found in Mohar [Moh].
3. Theorem (4.3.2) appears explicitly in [LZ], but as remarked also there, it is a corollary to the accumulation of the work of Cheeger, Buser, Dodziuk, Alon, Milman, Burger and last but not least, Brooks, who was the first to put all of this together. Burger [Bur3] gives a more precise connection between Property (τ) and λ_1.
4. Some of the details of Example 4.3.3C were provided to us by George Glauberman and Jim Lewis.

5. In our discussion of Theorem 4.3.2 we assume compactness for simplicity. But it works more generally: for example, if Γ is a lattice of a semi-simple group G, the theorem is also true with $M = X/\Gamma$ where X is the symmetric space associated with Γ (cf. [Br2] and the references therein). In fact, we used (4.3.2) in the proof of (4.4.2) even though $SL_2(\mathbb{Z})$ is not cocompact. It should be useful to formulate a more general theorem of this type, but we are not sure what are the natural general conditions.
6. A similar approach to our Theorem (4.4.3) is in Buser's paper [Bu3].

5 The Representation Theory of PGL_2

5.0 Introduction

Let $G_\infty = (PGL_2(\mathbb{R}))^0 = PSL_2(\mathbb{R})$, $G_p = PGL_2(\mathbb{Q}_p)$ where $\mathbb{Q}_p$ denotes the field of p-adic numbers and G will denote either G_∞ or G_p. Similarly, let $K_\infty = SO(2)/\{\pm 1\}$, $K_p = PGL_2(\mathbb{Z}_p)$ and K is either K_∞ or K_p. By P we denote the group of upper triangular matrices in G, $P = \{\begin{pmatrix} a & b \\ 0 & d \end{pmatrix} \in G\}$.

Let $\chi_s, s \in \mathbb{C}$ be the one-dimensional character of P, $\chi_s : P \to \mathbb{C}^*$, defined by:

$$\chi_s \begin{pmatrix} a & b \\ 0 & d \end{pmatrix} = |ad^{-1}|^s$$

where $|ad^{-1}|$ is the usual absolute value in $\mathbb{R}$, and the p-adic absolute value $|x| = p^{-v(x)}$ in case of $\mathbb{Q}_p$ ($v(x)$ denotes the valuation of x). By taking the induced representation $\operatorname{Ind}_P^G(\chi_s) = \rho_s$, one obtains an representation of G. It turns out that all irreducible unitary representations of G of class one (i.e., representations in which there exists a K-invariant vector) are obtained this way (after defining an appropriate inner product), for the following values of $s \in \mathbb{C}$:

Namely, the «class one» part of the unitary dual of G is a union of two sets: The principal series $= \{it | t \in \mathbb{R}\}$ and the complementary series that is equal to $\{s | -\frac{1}{2} \leq s \leq \frac{1}{2}\}$ in case of G_∞ and to $\{s + \frac{n\pi i}{\log p} | -\frac{1}{2} \leq s \leq \frac{1}{2}, n \in \mathbb{Z}\}$ in the case of G_p.

If ρ_s is such a representation, then there is an essentially unique K-invariant vector u, and the corresponding «matrix coefficient» $\varphi_s(g) = \langle \rho_s(g)u, u \rangle$ is a K-spherical function. In particular, it is bi-K-invariant and induces a function on G/K. In case of $G = G_p$, G/K is a $(p+1)$-regular tree to be described in Section 5.3 and for $G = G_\infty$, G/K is the upper half-plane. The function φ_s on G/K turns out to be an eigenfunction of the relevant Laplacian: In case of G_∞, it is of $\Delta = -y^2 \left(\frac{\partial^2}{\partial x^2} + \frac{\partial^2}{\partial y^2} \right)$ with eigenvalue $\lambda(s) = \frac{1}{4} - s^2$. In the case of G_p, let δ be the defined on the tree by: $(\delta f)(x) = \sum_{y \sim x} f(y)$ (the sum is over vertices adjacent to x). (So, $\delta = (p+1)I - \Delta$ where Δ is the combinatorial Laplacian defined in Section 4.2.) This time the corresponding eigenvalue of δ is: $\lambda(s) = p^{\frac{1}{2}}(p^s + p^{-s})$.

The connection between the representations and eigenvalues of the Laplacian (either Δ or δ) is also expressed in the following proposition:

Proposition 5.0.1. *Let Γ be a lattice in G. Then ρ_s appears as a subrepresentation of $L^2(\Gamma \backslash G)$ if and only if $\lambda(s)$ is an eigenvalue of Δ (resp: δ) on the surface (resp: graph) $\Gamma \backslash G/K$.*

Recall from Section 4.4 that Selberg's conjecture is that $\lambda(s) \geq \frac{1}{4}$ for congruence subgroups of the modular group. This exactly amounts to saying that $s = it$, i.e., no complementary series representations occur in $L^2(\Gamma\backslash G_\infty)$. The analog for G_p is:

Proposition 5.0.2. *Let $G = G_p$ and Γ be a lattice in G. Then no representation from the complementary series occurs in $L^2(\Gamma\backslash G)$ if and only if*

$$|\lambda(s)| = \left|p^{1/2}(p^s + p^{-s})\right| \leq 2\sqrt{p},$$

i.e., if and only if the graph $\Gamma\backslash G/K$ is a Ramanujan graph.

The fact that indeed there are Γ's for which this statement is valid (and hence will provide Ramanujan graphs) is a consequence of the validity of the classical Ramanujan-Peterson conjecture, however, this is shown in Chapters 6 and 7. Meanwhile, in the sections of this chapter, we will describe and essentially prove the results mentioned in the introduction. We begin with a presentation of the unitary dual of G. Though this subject is well known and is discussed in several books, we were unable to find a convenient reference for both cases G_∞ and G_p for readers who want to know only this and not more. We therefore present a virtually self-contained exposition, i.e., one which uses results from the literature but only if they are easily accessible with proofs. We begin in Section 5.1 by establishing the connection between representations and spherical functions. Here we follow Chapter IV of Lang's book «$SL_2(\mathbb{R})$». (The reader is encouraged to read its 14 pages in full – it is independent from the rest of the book.) Once we have this, we can present in Section 5.2 the representations of class one of $PSL_2(\mathbb{R})$ (i.e., the principal and complementary series; we ignore the discrete series which is not so important for our purposes). Here we partially use notes of H. Furstenberg's lectures in a Hebrew University seminar in 1985. In Section 5.3 we describe (following Serre [S1]) the tree associated with $PGL_2(\mathbb{Q}_p)$, and in Section 5.4 we use Section 5.3 to describe the principal and complementary series of representations of G_p. In Section 5.5 we discuss the connection between the eigenvalues of $\Gamma\backslash G/K$ and the representations appearing in $L^2(\Gamma\backslash G)$.

At that point the analogy between Selberg's conjecture and Ramanujan graphs becomes clear.

5.1 Representations and spherical functions

In this section, G will be one of the groups $PSL_2(\mathbb{R})$, $GL_2(\mathbb{R})$, $PGL_2(\mathbb{Q}_p)$ or $GL_2(\mathbb{Q}_p)$, and K a maximal compact subgroup of G, i.e., K is either $PSO(2)$, $O(2)$, $PGL_2(\mathbb{Z}_p)$ or $GL_2(\mathbb{Z}_p)$, respectively.

Let (H, ρ) be an irreducible unitary representation of G on a Hilbert space H and $0 \neq u \in H$ any vector. In Chapter 3 we introduced the coefficient function $\phi : G \to \mathbb{C}$, $\phi(g) = \langle\rho(g)u, u\rangle$.

Lemma 5.1.1. *ϕ determines ρ, i.e., there is a unique irreducible representation with this ϕ as a coefficient function.*

Proof. Given $g \in G$, then $\rho(g)$ is determined once we know $\langle \rho(g)v_1, v_2 \rangle$ for every v_1, v_2 in H. In fact, it suffices to know it on a dense subset of H. Such a dense subset is obtained by taking sums of the form $v_1 = \sum c_i \rho(g_i)u$, since ρ is irreducible. Now if $v_2 = \sum d_j \rho(g_j)u$, then

$$\langle \rho(g)v_1, v_2 \rangle = \sum c_i \overline{d_j} \phi(g_j^{-1} g_i),$$

and thus $\rho(g)$ is determined. □

If the vector u is K-invariant, i.e., $\rho(k)u = u$ for every $k \in K$, then the corresponding ϕ is bi-K-invariant, i.e., $\phi(k_1 g k_2) = \phi(g)$ for every $k_1, k_2 \in K$, and $g \in G$ (since $\phi(k_1 g k_2) = \langle \rho(k_1 g k_2)u, u \rangle = \langle \rho(g)\rho(k_2)u, \rho(k_1^{-1})u \rangle = \langle \rho(g)u, u \rangle = \phi(g)$).

Let $C = C_c(G//K)$ denote the set of bi-K-invariant (bi-invariant for short) functions on G with compact support. C is an algebra with respect to convolution

$$f_1 * f_2(x) = \int_G f_1(xg^{-1}) f_2(g) dg.$$

Proposition 5.1.2. (cf. [La], page 53) *$C_c(G//K)$ is a commutative algebra. If (H, ρ) is a unitary representation of G, ρ induces a representation $\overline{\rho}$ of the algebra $C_c(G//K)$ by:*

$$\overline{\rho}(f) = \int_G f(g)\rho(g) dg \quad \textit{for} \quad f \in C.$$

Let H^K denote the space of K-invariant vectors in H; then $\overline{\rho}(f)(H^K) \subseteq H^K$ and hence $(H^K, \overline{\rho})$ is a representation of C.

Proposition 5.1.3. (cf. [La], page 54) *Assume $H^K \neq \{0\}$ and ρ is irreducible. Then, $(H^K, \overline{\rho})$ is irreducible.*

From (5.1.2), (5.1.3) and Schur's Lemma one deduces:

Proposition 5.1.4. (cf. [La], page 55) *Let (H, ρ) be a unitary irreducible representation of G; then $\dim H^K = 1$ or $\dim H^K = 0$.*

In the first case, $\dim H^K = 1$, we say that ρ is a representation of class one. *In what follows we will be interested only in such representations.* G has representations with $H^K = \{0\}$. (See [La] or [GGPS].) Such representations are «far away» from the trivial representation in the Fell topology of the unitary dual since they do not contain almost invariant vectors: if v is ε-invariant with respect to K (i.e., $\|\rho(k)v - v\| < \varepsilon$ for every $k \in K$) and $\varepsilon < 1$, then $\int_K \rho(k)v dk$ is in

H^K and it is nonzero – a contradiction. This shows that such representations are uniformly bounded away from the trivial representation.

So, assume now that (H, ρ) is an irreducible unitary representation and $\dim H^K = 1$. Let $u \in H^K$ with $\|u\| = 1$ and $\varphi : G \to \mathbb{C}$ the function $\varphi(g) = \langle \rho(g)u, u \rangle$. Note that φ is bi-K-invariant. It, in fact, satisfies also the other conditions of the following definition:

Definition 5.1.5. *A function f on G is called K-spherical (or spherical) if the following are satisfied:*

1. *f is bi-invariant and continuous;*
2. *$f(e) = 1$ where e is the unit element of G; and*
3. *f is an eigenfunction of $C_c(G//K)$ on the right, i.e., $f * \psi = \lambda(f, \psi) f$ for $\psi \in C_c(G//K)$ and some complex number $\lambda(f, \psi)$.*

Proposition 5.1.6. (cf. [La], page 61) *φ as above is a spherical function. Moreover, φ is also positive definite, i.e., for every $x_1, \dots, x_n \in G$ and $a_1, \dots, a_n \in \mathbb{C}$ we have:*

$$\sum_{i,j} \varphi(x_i^{-1} x_j) a_i \overline{a_j} \geq 0.$$

This is because:

$$\sum_{i,j} \varphi(x_i^{-1} x_j) a_i \overline{a_j} = \sum_{i,j} \langle \varphi(x_i^{-1} x_j) a_i \overline{a_j} u, u \rangle$$
$$= \langle \sum_j \rho(x_j) \overline{a_j} u, \sum_i \rho(x_i) \overline{a_i} u \rangle \geq 0.$$

In light of Lemma 5.1.1 and Proposition 5.1.6, the following is not surprising.

Theorem 5.1.7. ([La], page 65) *The above association $(H, \rho) \mapsto \varphi$ is a bijection from the set of irreducible unitary representations of G having K-fixed vectors to the positive definite spherical functions on G.*

So, our goal in classifying the irreducible unitary representations of class one will be first to find the spherical functions. Secondly, we will have to decide which of them are positive definite or equivalently come from unitary representations.

The following proposition gives a useful method to find spherical functions:

Proposition 5.1.8. *Let P be a closed subgroup of G and assume $G = PK$. Let χ be a character (i.e., a continuous homomorphism) $\chi : P \to \mathbb{C}^*$ which is trivial on $P \cap K$. Extend this to a function on G by setting $\chi(pk) = \chi(p)$. Then χ is a right eigenvector of $C_c(G//K)$, namely*

$$\chi * \psi(x) = \lambda(\chi, \psi) \chi(x),$$

and a function f such that $f(x) = \int_K \chi(kx) dk$ is K-spherical with eigenvalues $\lambda(\chi, \psi) = \chi * \psi(e) = \lambda(f, \psi)$.

Proof. This is proved in [La, Theorem 4, page 55] under the assumption that $P \cap K = \{1\}$, but the proof works word by word for the somewhat more general case. $\square$

Before turning to concrete examples we record the following easy property:

Proposition 5.1.9. *If (H, φ) is an irreducible unitary representation, $u \in H^K$ with $\|u\| = 1$ and $\varphi(g) = \langle \rho(g)u, u\rangle$, then for every $g \in G$, $\varphi(g)$ is real and $|\varphi(g)| \leq \varphi(e) = 1$.*

Proof. First note that $\varphi(g^{-1}) = \langle \rho(g^{-1})u, u\rangle = \langle u, \rho(g)u\rangle = \overline{\varphi(g)}$. Second, in all possibilities of G one can write g^{-1} as kgk' for some $k', k \in K$ (check!); hence $\varphi(g^{-1}) = \varphi(g)$ and so $\varphi(g)$ is real. Moreover, $|\varphi(g)| = |\langle \rho(g)u, u\rangle| \leq \|\rho(g)(u)\| \, \|u\| = 1$ (since ρ is unitary). $\square$

5.2 Irreducible representations of $PSL_2(\mathbb{R})$ and eigenvalues of the Laplacian

In this section we specialize to the case $G = PSL_2(\mathbb{R})$ and $K = PSO(2)$. The symmetric space G/K can be identified with the upper half-plane $\mathbb{H} = \{x + iy | x, y \in \mathbb{R}, y > 0\}$ via the action of G on $\mathbb{H}$, $g(z) = \frac{az+b}{cz+d}$ for $g = \begin{pmatrix} a & b \\ c & d \end{pmatrix} \in G$. So, the stabilizer of $i \in \mathbb{H}$ is $PSO(2)$. We introduced in Chapter 4 the Laplacian Δ of $\mathbb{H}$, which is defined as $\Delta = -y^2 \left(\frac{\partial^2}{\partial x^2} + \frac{\partial^2}{\partial y^2} \right)$, and it commutes with the action of G. Another (isomorphic) model for G/K is the unit disc $U = \{z \in \mathbb{C} | \, |z| < 1\}$ via the action

$$\begin{pmatrix} a & b \\ c & d \end{pmatrix}(w) = \frac{[(a+d) + (b-c)i]w + [(a-d) + (-b-c)i]}{[(a-d) + (b+c)i]w + [(a+d) + (-b+c)i]}.$$

Indeed, the transformation

$$\begin{pmatrix} a & b \\ c & d \end{pmatrix} \mapsto \frac{1}{2i} \begin{pmatrix} 1 & -i \\ 1 & i \end{pmatrix} \begin{pmatrix} a & b \\ c & d \end{pmatrix} \begin{pmatrix} i & i \\ -1 & 1 \end{pmatrix}$$

conjugates $SL_2(\mathbb{R})$ to $SU(1,1)$. This time K is the stabilizer of 0 and it acts on U by rotations. The Laplacian here gets the form of

$$\Delta = -(1 - x^2 - y^2) \left(\frac{\partial^2}{\partial x^2} + \frac{\partial^2}{\partial y^2} \right).$$

It will be convenient to keep in mind both models.

A spherical function φ being K-invariant from the right can be thought as a function on $\mathbb{H}$. Moreover, being K-invariant from the left as well, φ is constant on K-orbits on $\mathbb{H}$, or what is easier to imagine, a function on U which is constant on circles around 0, i.e., a «radial function». Thus φ has the same derivative in all directions and one can also consider ϕ as a function of one variable. From [La, Theorem 8, p. 199] it follows that every spherical function φ is an eigenfunction of the Laplacian with respect to some eigenvalue $\lambda \in \mathbb{C}$. Thinking of φ as a function on U we have the following information:

$$\begin{cases} \Delta\varphi = \lambda\varphi & \text{as was just said} \\ \varphi(0) = 1 & \text{since } \varphi(e) = 1 \\ \varphi'(0) = 0 & \text{since } \varphi \text{ is radial.} \end{cases}$$

The equation $\Delta\varphi = \lambda\varphi$ is a second-order differential equation, and with the two other conditions we can deduce from the fundamental theorem of differential equations that φ is uniquely determined. Let φ^λ be, therefore, the unique spherical function satisfying $\Delta\varphi^\lambda = \lambda\varphi^\lambda$ (λ now is an arbitrary but fixed complex number).

Now, if φ is positive definite, i.e., comes from a unitary representation, then we saw (Proposition 5.1.9) that φ is real and 0 is a maximum point, which implies that $\varphi'(0) = 0$ and $\varphi''(0) < 0$ so $\frac{\partial^2}{\partial x^2}\varphi\Big|_0 = \frac{\partial^2}{\partial y^2}\varphi\Big|_0 \leq 0$. Thus if φ^λ is also positive definite then $\lambda\varphi^\lambda(0) = \Delta(\varphi^\lambda)(0) = -(1-0^2-0^2)\left(\frac{\partial^2}{\partial x^2} + \frac{\partial^2}{\partial y^2}\right)(\varphi)(0) \geq 0$ (and in particular λ is real). We therefore conclude:

Theorem 5.2.1. *If (H, ρ) is an irreducible unitary representation of $G = PSL_2(\mathbb{R})$ of class one with an associated spherical function φ, then $\Delta\varphi = \lambda\varphi$ for some nonnegative real number λ. For every $\lambda \geq 0$ there is a unique such spherical function φ^λ satisfying $\Delta(\varphi^\lambda) = \lambda\varphi^\lambda$.*

Our next goal will be to show that indeed for every $\lambda \geq 0$, such a spherical function comes from a representation. To this end we use Proposition 5.1.8. Let $P = \{\begin{pmatrix} a & b \\ 0 & a^{-1} \end{pmatrix} \in PSL_2(\mathbb{R})\}$. From the Gram-Schmidt orthonormalization theorem it follows that $G = PK$. Characters χ on P are given by: Let $t \in \mathbb{C}$ and $\chi_t\begin{pmatrix} a & b \\ 0 & a^{-1} \end{pmatrix} = a^{2t}$. As a function on $\mathbb{H}$, χ_t gives: $\chi_t(x+yi) = y^t$ since $\begin{pmatrix} a & b \\ 0 & a^{-1} \end{pmatrix}(i) = a^2 i + ba$, so $\begin{pmatrix} a & b \\ 0 & a^{-1} \end{pmatrix}$ is associated with the point $x + iy = ba + ia^2$. According to the procedure in Proposition 5.1.8, the function

$$\varphi_t(z) = \int_K (k \cdot \chi_t)(z)dk = \int_K \chi_t(kz)dk = \int_K \operatorname{imag}(kz)^t dk$$

is a spherical function.

We want to find $\lambda = \lambda(t)$, the corresponding eigenvalue of $\Delta, \Delta\varphi_t = \lambda(t)\varphi_t$.

Let's compute:

$$\Delta(\chi_t) = -y^2 \left(\frac{\partial^2}{\partial x^2} + \frac{\partial^2}{\partial y^2} \right) y^t$$
$$= t(1-t)y^t = t(1-t)\chi_t.$$

As Δ commutes with the action of k:

$$\Delta(\varphi_t) = \Delta\left(\int_K (k \cdot \chi_t) dk \right) = \int_K (k \cdot \Delta(\chi_t)) dk$$
$$= \int_K k \cdot (t(1-t)\chi_t) dk = t(1-t) \int_K k\chi_t dk = t(1-t)\varphi_t.$$

The conditions λ is real and $\lambda \geq 0$ lead to:

Theorem 5.2.2. *The spherical functions φ_t as above, satisfying $\Delta\varphi_t = t(1-t)\varphi_t$ and $\lambda = t(1-t) \geq 0$, are from two sources:*

1. *The principal series: $t \in \{\frac{1}{2} + ir | r \in \mathbb{R}\}$ and then $\lambda \geq \frac{1}{4}$.*
2. *The complementary series: $t \in [0, 1]$ and $0 \leq \lambda \leq \frac{1}{4}$.*

To get actual representations, we argue as follows (following [La, Chapter III]): Let

$$\chi_s : P \to \mathbb{C}^* \quad \text{be the character}$$

$$\chi_s \begin{pmatrix} a & b \\ 0 & a^{-1} \end{pmatrix} = a^{2s}$$

and let $\rho_s = \mathrm{Ind}_P^G(\chi_s)$. The discussion in [La, Chapter III, Section 3] shows that ρ_s has a K-fixed vector and the associated spherical function is $\varphi_{s+\frac{1}{2}}$ in the previous notation.

Moreover, if $s = ir$, then χ_s is a unitary character, and hence ρ_s is also unitary. These observations and Theorem 5.2.2 almost complete the proof of the following:

Theorem 5.2.3. *The irreducible unitary representations of $PSL_2(\mathbb{R})$ of class one (i.e., with K-invariant vector) are given by the following parameterization:*

1. *The principal series: ρ_s for $s \in \{ir | r \in \mathbb{R}\}$ with spherical function*

$$\varphi_{\frac{1}{2}+s} \quad \text{and} \quad \lambda = \frac{1}{4} - s^2 \geq \frac{1}{4}.$$

2. *The complementary series: ρ_s for $-\frac{1}{2} \leq s \leq \frac{1}{2}$ with spherical function*

$$\varphi_{\frac{1}{2}+s} \quad \text{and} \quad 0 \leq \lambda = \frac{1}{4} - s^2 \leq \frac{1}{4}.$$

The part of the proof which is still missing is that indeed the representations of the complementary series really exist in the *unitary* dual. As it is presented now they are not unitary. But, it is known that by changing the inner product they can be unitarized. For details see [La] or [GGPS]. We should also mention that the above parameterization is redundant: the representations ρ_s and ρ_{-s} are equivalent, as can be seen from the fact that $\lambda = \frac{1}{4} - s^2$ and that λ uniquely determines the spherical function and hence also the representation.

Finally, we mention that all the principal series representations are bounded away uniformly from the trivial representation in the Fell topology of the unitary dual. To see this, notice first (as in Section 5.1) that a $\overline{K}$-almost-invariant-vector for a compact set $\overline{K}$ containing K gives by averaging over K a $\overline{K}$-almost-invariant-vector which is K-invariant. But we said that the K-invariant vectors form a one-dimensional space generated, say, by u. So we have to check only the invariance properties of u. If $g \in G$ is fixed and $g \notin K$, then $\|\rho(g)u - u\|^2 = 2(1 - \langle\rho(g)u, u\rangle) = 2(1 - \varphi(g))$ where φ is the associated spherical function. So if $\rho = \rho_s$ for $s = ir$, then

$$\varphi_{\frac{1}{2}+ir}(g) = \int_K \operatorname{imag}\left(kg(i)\right)^{\frac{1}{2}+ir} dk.$$

This last expression is bounded away from 1 independently of r. In fact, the parameterization of the class one unitary dual of G_∞ as a subset of $\mathbb{C}$ reflects also the entire topology. In particular the complementary series goes to the trivial representation (which is represented by $\rho \pm \frac{1}{2}$), and $PSL_2(\mathbb{R})$ does not have property (T) as was already proven in (3.1.9). This is also the *only* way for representations to converge to the trivial one, as we already have shown that those with no K-invariant vectors are also bounded away from the trivial representation.

5.3 The tree associated with $PGL_2(\mathbb{Q}_p)$

In this section we present the «symmetric space» of $PGL_2(\mathbb{Q}_p)$, which will be a $(p+1)$-regular tree X. Just like the upper half-plane $\mathbb{H}$ which is a quotient of $G = PSL_2(\mathbb{R})$ by its maximal compact subgroup $K = PSO(2)$, X is the quotient $PGL_2(\mathbb{Q}_p)/PGL_2(\mathbb{Z}_p)$ (or more precisely: this coset space is the set of vertices of X). This tree is the Bruhat-Tits building associated with the p-adic Lie group $PGL_2(\mathbb{Q}_p)$, but following [S1] we will give a more concrete construction of X.

Let $V = \mathbb{Q}_p \times \mathbb{Q}_p$ be the two-dimensional vector space over the field of p-adic numbers. A $\mathbb{Z}_p$-lattice in V is a $\mathbb{Z}_p$-submodule L of V generated by two linearly independent vectors in V. The standard lattice L_0 is the one generated by the standard basis e_1 and e_2, i.e., $L_0 = \mathbb{Z}_p \times \mathbb{Z}_p$. Two lattices L_1 and L_2 are said to be equivalent if there exists $0 \neq \alpha \in \mathbb{Q}_p$ such that $L_2 = \alpha L_1$. We define a graph X as follows: Its set of vertices will be the set of equivalence classes of lattices. Two equivalence classes $[L_1]$ and $[L_2]$ are said to be adjacent on the graph if there are representatives $L_i' \in [L_i]$ such that $L_1' \subseteq L_2'$ and $[L_2' : L_1'] = p$. (This is a symmetric relation since pL_2' is equivalent to L_2' and $pL_2' \subseteq L_1'$ with $[L_1' : pL_2'] = p$.) This defines a structure of a graph.

The group $GL_2(\mathbb{Q}_p)$ acts transitively on the set of $\mathbb{Z}_p$-lattices and its center preserves equivalent classes. Thus, $PGL_2(\mathbb{Q}_p)$ acts transitively on the vertices of X. The stabilizer of the equivalence class of the standard lattice L_0 is $PGL_2(\mathbb{Z}_p)$. The set of vertices can be, therefore, identified with $PGL_2(\mathbb{Q}_p)/PGL_2(\mathbb{Z}_p)$. The action of $G = PGL_2(\mathbb{Q}_p)$ clearly preserves the relation of adjacency and so G acts as a transitive group of automorphisms of X. This implies that X is a k-regular graph for some k. Checking the possible neighbors of the (equivalence class of the) standard lattice, we easily see that L_0 has exactly $p+1$ sublattices of index p (just as the number of the one dimensional subspaces of $L_0/pL_0 \simeq \mathbb{F}_p \times \mathbb{F}_p$, which is the number of points on the projective line over the finite field $\mathbb{F}_p$ of p elements). Thus X is a $(p+1)$-regular graph. One way to present the neighbors of L_0 is by looking at the matrices:

$$A_i = \begin{pmatrix} p & i \\ 0 & 1 \end{pmatrix} \quad \text{for} \quad i = 0, \ldots, p-1 \quad \text{and} \quad A_\infty = \begin{pmatrix} 1 & 0 \\ 0 & p \end{pmatrix}.$$

Then $A_0(L_0), \ldots, A_{p-1}(L_0), A_\infty(L_0)$ are $p+1$ representatives of the $p+1$ vertices adjacent to the vertex $[L_0]$.

X is indeed a connected tree. A proof is given in [S1, p. 70]. A quick proof is the following: It suffices to show that there is a *unique* path from $x_0 = [L_0]$ to any other vertex x. A representative L of x can be picked up inside $L_0 = \mathbb{Z}_p \times \mathbb{Z}_p$ and, moreover, L can be chosen in such a way that L_0/L is a cyclic group (otherwise replace L by $p^{-1}L$). The existence and uniqueness of a path from L_0 to L follows now from the existence and uniqueness of the Jordan-Hölder series in a finite cyclic p-group.

We identified X with G/K, so functions on X can be thought as right K-invariant functions on G and vice versa. K acts on $X = G/K$ by fixing $x_0 = [L_0]$ and acting transitively on the set of vertices of distance n from L_0 (for every n). Hence, bi-K-invariant functions on G correspond to radial functions on X, i.e., functions $f(x)$ in which $f(x)$ depends only on the distance $d(x, x_0)$ from x to x_0. The algebra $C_c(G//K)$, introduced in Section 5.1, consists of functions of this type for which $f(x) = 0$ for almost all x. An example of such a function is the characteristic function $\overline{\delta}$ of the double coset $K \begin{pmatrix} p & 0 \\ 0 & 1 \end{pmatrix} K$. This function gives the value 1 to the vertices of distance 1 from $x_0 = [L_0]$ and 0 elsewhere. $\overline{\delta}$ defines an operator δ, called the *Hecke operator* on the functions on X, by:

$$\delta(f) = \delta(t) = f * \overline{\delta}.$$

(In order that this definition, which uses convolution, will make sense, we think of $\overline{\delta}$ and f as functions on G, bi-invariant and right-invariant respectively.) Spelling out this definition to f as a function on X, $\delta(f)$ is the following:

$$\delta(f) = \sum_{d(y,x)=1} f(y).$$

So, δ is $(p+1)$ times the *averaging operator*. Recall from Chapter 4 the definition of the combinatorial Laplacian Δ. So, $\delta = (p+1)Id - \Delta$. (Sometimes, δ itself is called the Laplacian.)

In the next section we will use this information to find spherical functions and hence representations of $G = PGL_2(\mathbb{Q}_p)$.

5.4 Irreducible representations of $PGL_2(\mathbb{Q}_p)$ and eigenvalues of the Hecke operator

In this section we present the analog of Section 5.2 for $G = PGL_2(\mathbb{Q}_p)$ instead of $PSL_2(\mathbb{R})$. (The reason we work with $PGL_2(\mathbb{Q}_p)$ and not $PSL_2(\mathbb{Q}_p)$ is just a matter of convenience – the second does not act transitively on the tree – but everything can be done similarly for $PSL_2(\mathbb{Q}_p)$.) We want to find the unitary irreducible representations of G of class one (i.e., those with K-fixed vectors). By Section 5.1 this amounts to find the positive definite spherical functions. Just like in Section 5.2, every spherical function is an eigenfunction of the Laplacian (or equivalently the Hecke operator). Here this is even simpler. The function $\overline{\delta}$ defined in Section 5.3 (i.e., the characteristic function of the double coset $K \begin{pmatrix} p & 0 \\ 0 & 1 \end{pmatrix} K$) is in $C_c(G//K)$ so every spherical function φ is by Definition (see 5.1.5) an eigenfunction of convolving with $\overline{\delta}$ from the right, i.e., $\delta(\varphi) = \lambda\varphi$ for some $\lambda \in \mathbb{C}$. Such a spherical function φ (being bi-K-invariant) is, as explained in Section 5.3, given by a sequence of numbers $a_0 = 1, \ldots, a_n, \ldots$ such that $\varphi(x) = a_n$ if $d(x, x_0) = n$. The equation $\delta(\varphi) = \lambda\varphi$ spells out as the following recursive relation:

$$\begin{cases} \lambda a_n = a_{n-1} + p a_{n+1} & \text{for } n \geq 1 \\ \lambda a_0 = (p+1) a_1 & . \end{cases}$$

We deduce that the sequence is uniquely determined by the eigenvalue λ, and in fact it is easy to compute it recursively, e.g., $a_1 = \frac{\lambda}{p+1}$, $a_2 = \frac{\lambda^2-(p+1)}{p(p+1)}$, etc.

So for every λ we have a unique spherical function φ^λ. But, on those coming from unitary representations we have extra conditions (see (5.1.9)): φ^λ should be real and gets its maximum at e. From the equation $a_1 = \frac{\lambda}{p+1}$ we deduce that λ should be real and $|\lambda| \leq p+1$. Summarizing this we have (in analogy with Theorem (5.2.1)):

Theorem 5.4.1. *If (H, ρ) is an irreducible unitary representation of $G = PGL_2(\mathbb{Q}_p)$ of class one with an associated spherical function φ, then $\delta(\varphi) = \lambda\varphi$ for some real number λ, $|\lambda| \leq p+1$. For every such λ there is a unique spherical function φ^λ satisfying $\delta(\varphi^\lambda) = \lambda\varphi^\lambda$.*

Our next goal will be to show that indeed for every $|\lambda| \leq p+1$ such a spherical function comes from a representation. To this end we use (5.1.8).

Let $P = \{\begin{pmatrix} a & b \\ 0 & d \end{pmatrix} \in PGL_2(\mathbb{Q}_p)\}$. Then $G = PK$, with $P \cap K = \{\begin{pmatrix} a & b \\ 0 & d \end{pmatrix} \Big| a, d \in \mathbb{Z}_p^*, b \in \mathbb{Z}_p\}$. The characters $\chi_t : P \to \mathbb{C}$ given by:

$$\chi_t \begin{pmatrix} a & b \\ 0 & d \end{pmatrix} = |ad^{-1}|^t$$

satisfy $\chi_t(P \cap K) = 1$. So by (5.1.8), $\varphi_t(x) = \int_K \chi_t(kx)dk$ is a K-spherical function (where χ_t is extended to whole G by $\chi(pk) = \chi(p)$).

To compute the associated eigenvalue $\lambda = \lambda(t)$ of δ, it suffices to compute $a_1 = \varphi_t(x_1)$ for some vertex x_1 of distance one from $x_0 = [L_0]$ and then $\lambda = (p+1)a_1$. So, let $g = \begin{pmatrix} p & 0 \\ 0 & 1 \end{pmatrix}$ and so,

$$\begin{aligned} a_1 = \varphi_t \begin{pmatrix} p & 0 \\ 0 & 1 \end{pmatrix} &= \int_K \chi_t(kg)dk \\ &= \sum_{i=0}^{p-1} \frac{1}{p+1}\chi_t \begin{pmatrix} p & i \\ 0 & 1 \end{pmatrix} + \frac{1}{p+1}\chi_t \begin{pmatrix} 1 & 0 \\ 0 & p \end{pmatrix} = \frac{1}{p+1}(p \cdot p^{-t} + p^t). \end{aligned}$$

Thus $\lambda = \lambda(t) = p^{1-t} + p^t$. To ensure that λ will be real; first suppose that p^t is real, in which case t is of the form $r + \frac{n\pi i}{\log p}$ where r is real between zero and one since $|\lambda| \leq p+1$, and n is any integer. If p^t is not real, then for $p^{1-t} + p^t$ to be real we must have that, if $t = x + iy$ then: $p^{1-x}p^{-iy} + p^x p^{iy} \in \mathbb{R}$ and so $p^{1-x} = p^x$, which implies $x = \frac{1}{2}$. So $t = \frac{1}{2} + ir$ for $r \in \mathbb{R}$.

To summarize:

Theorem 5.4.2. *The spherical functions φ_t as above such that $\delta(\varphi_t) = (p^{1-t} + p^t)\varphi_t$ and $\lambda = p^{1-t} + p^t$ is real of absolute value $\leq p+1$ come from two sources:*

1. *The principal series: $t \in \{\frac{1}{2} + ir | r \in \mathbb{R}\}$ and then $|\lambda| \leq 2\sqrt{p}$.*
2. *The complementary series: $t \in \{r + \frac{n\pi i}{\log p} | 0 \leq r \leq 1, n \in \mathbb{Z}\}$ and then $2\sqrt{p} \leq |\lambda| \leq p+1$.*

To get actual representations we argue as follows (following [La, Chapter III]) and in analogy to Section 5.2. Let $\chi_s : P \to \mathbb{C}^*$ be the character

$$\chi_s \begin{pmatrix} a & b \\ 0 & d \end{pmatrix} = |ad^{-1}|^s$$

and let $\rho_s = \mathrm{Ind}_P^G(\chi_s)$.

The discussion in [La, Chapter III, Section 3] shows that ρ_s has a K-fixed vector and that the associated spherical function is $\varphi_{s+\frac{1}{2}}$ in the previous notation. Moreover, if $s = ir$, then χ_s is a unitary character and so ρ_s is also unitary.

Theorem 5.4.3. *The irreducible unitary representations of $PGL_2(\mathbb{Q}_p)$ of class one (i.e., with K-invariant vectors) are given by the following parametrization:*

1. *The principal series: ρ_s for $s \in \{ir | r \in R\}$ with spherical function $\varphi_{\frac{1}{2}+s}$ and $\lambda = p^{\frac{1}{2}-s} + p^{\frac{1}{2}+s}$ so $|\lambda| \leq 2\sqrt{p}$.*
2. *The complementary series: ρ_s for $s \in \left[-\frac{1}{2}, \frac{1}{2}\right] + \mathbb{Z}\frac{\pi i}{\log p}$, with spherical function $\varphi_{\frac{1}{2}+s}$ and $\lambda = p^{\frac{1}{2}}(p^{-s} + p^s)$, so $2\sqrt{p} \leq |\lambda| \leq p+1$.*

The part of proof which is missing is again the unitarization of the complementary series. For this see [GGPS]. Just as for $PSL_2(\mathbb{R})$, there are repetitions in this parameterization, but this time even more significant ones: s, $-s$ and $s + \mathbb{Z}\frac{2\pi i}{\log p}$ all give the same λ and hence the same spherical function and thus the same representation.

The principal series representations are also uniformly bounded away from the trivial representation since $\varphi_t\left(\begin{pmatrix} p & 0 \\ 0 & 1 \end{pmatrix}\right) = \frac{\lambda}{p+1} \ll 1$. (See the corresponding argument for $PSL_2(\mathbb{R})$.) Here it is not difficult to show that when $\lambda \to (p+1)$ or equivalently $s \to \frac{1}{2}$, the representations associated to ρ_λ must go to the trivial representation since the corresponding spherical functions go uniformly on compact sets to the constant function 1. Hence $PGL_2(\mathbb{Q}_p)$ does not have property (T).

Note also that for $s = \frac{1}{2} + \frac{\pi i}{\log p}$ (as well as for $s = \frac{1}{2} + (2n+1)\frac{\pi i}{\log p}$) the corresponding representation is $sg(g)$ (see [GGPS, p. 132]; this is the unique one-dimensional non-trivial representation of $PGL_2(\mathbb{Q}_p)$ which is trivial on $PGL_2(\mathbb{Z}_p)$; namely, the sign representation.)

5.5 Spectral decomposition of $\Gamma \backslash G$

In the previous section we saw the close connection between representations and eigenvalues of the Laplacian (resp: the Hecke operator). If λ is such an eigenvalue (where $\lambda \geq 0$ in case of G_∞ and $-(p+1) \leq \lambda \leq p+1$ in the case of G_p), we have a unique unitary representation ρ^λ of G such that the corresponding spherical function φ^λ satisfies $\Delta\varphi^\lambda = \lambda\varphi^\lambda$ (resp: $\delta\varphi^\lambda = \lambda\varphi^\lambda$). Note that there are several values of s such that $\rho_s = \rho^\lambda$ (in fact, any s with $\lambda = \frac{1}{4} - s^2$ in the case of G_∞ or $\lambda = p^{\frac{1}{2}}(p^s + p^{-s})$ in the case of G_p), but only one ρ^λ. In this section we will prove the following theorem:

Theorem 5.5.1 *Let Γ be a lattice in G_∞ (resp: G_p). Then λ is an eigenvalue of the Laplacian Δ (resp: the Hecke operator δ) acting on $\Gamma \backslash G/K$ if and only if ρ^λ occurs as a subrepresentation of $L^2(\Gamma \backslash G)$.*

Recall that we mentioned in Chapter 4 the conjecture of Selberg that for congruence subgroups Γ of $PSL_2(\mathbb{Z})$ in $G = PSL_2(\mathbb{R})$, $\lambda_1(\Gamma\backslash G/K) \geq \frac{1}{4}$. Theorems (5.2.3) and (5.5.1) now yield.

Corollary 5.5.2. *The Selberg conjecture holds, i.e.,* $\lambda_1(\Gamma\backslash G/K) \geq \frac{1}{4}$, *if and only if no complementary series representation is a subrepresentation of* $L^2(\Gamma\backslash G)$ *(except of the trivial representation which corresponds to* $\lambda_0 = 0$*).*

Now if Γ is a lattice in $PGL_2(\mathbb{Q}_p)$, then G/K is a $(p+1)$-regular tree and $\Gamma\backslash G/K$ is a $(p+1)$-regular graph (in fact a finite graph, since every lattice in $PGL_2(\mathbb{Q}_p)$ is cocompact – see [S1, p. 84]). In Chapter 4 we called a k-regular graph a Ramanujan graph if, with the exception of $\pm k$, all eigenvalues of $\delta = kI - \Delta$ ($\Delta =$ the combinatorial Laplacian) are of absolute value $\leq 2\sqrt{k-1}$. Specializing to our case $k = p+1$ we have:

Corollary 5.5.3. *If* Γ *is a (cocompact) lattice in* $G = PGL_2(\mathbb{Q}_p)$ *then* $\Gamma\backslash G/K$ *is a Ramanujan graph if and only if no complementary series representation occurs in* $L^2(\Gamma\backslash G)$ *(except for the trivial one which corresponds to* $\lambda = p+1$, *and possibly the sign representation* $sg(g)$ *which corresponds to* $\lambda = -(p+1)$*).*

Remark. The representation sg did not come up in $PGL_2(\mathbb{R})$ and does not exist for $PSL_2(\mathbb{R})$ since every one-dimensional representation of $PGL_2(\mathbb{R})$ (or $PSL_2(\mathbb{R})$) which is trivial on K is trivial. This is not the case with $\mathbb{Q}_p$ since $|\mathbb{Q}_p^*/(\mathbb{Q}_p^*)^2| = 4$ while $|\mathbb{R}^*/(\mathbb{R}^*)^2| = 2$, whence $|PGL_2(\mathbb{Q}_p)/PSL_2(\mathbb{Q}_p)| = 4$ while $|PGL_2(\mathbb{R})/PSL_2(\mathbb{R})| = 2$. So, except for this inessential difference we see that the «Ramanujan graph property» is the precise analog of Selberg's conjecture. In the next chapter we will see that indeed it holds for congruence subgroups of arithmetic groups in $PGL_2(\mathbb{Q}_p)$. Thus, it will provide us with a source of Ramanujan graphs.

Proof of Theorem 5.5.1. We begin with proving the theorem for G_p.

Assume λ is an eigenvalue of δ on $\Gamma\backslash G/K$ with an eigenfunction f, with $\|f\| = 1$. We consider f as a function of G which is left invariant under Γ (and hence in $L^2(\Gamma\backslash G)$) and K-invariant from the right. f is therefore a K-invariant vector. To prove that the representation ρ^λ occurs in $L^2(\Gamma\backslash G)$, it suffices to compute the K-spherical function associated with f, i.e., $\varphi(g) = \langle\rho(g)\cdot f, f\rangle$, when ρ is the (right) action of G on $L^2(\Gamma\backslash G)$.

Lemma 5.5.4. *For every* $x, y \in G$, $\int_K f(xky)dk = \varphi^\lambda(y)f(x)$ *where* φ^λ *is the spherical function associated with the representation* ρ^λ.

Proof. Let $F_x(y) = \int_K f(xky)dk$; then F_x is a bi-K-invariant function (check!). We will show that it is also an eigenfunction of δ. Recall that δ commutes with the action π of G on functions on G/K given by $\pi(g)f(y) = f(g^{-1}y))$, and hence:

$$\delta(F_x(y)) = \delta\left(\int_K \pi(xk)^{-1}f(y)dk\right) = \int_K \pi(xk)^{-1}(\delta f)(y)dk$$
$$= \lambda \int_K \pi(xk)^{-1}f(y)dk = \lambda F_x(y).$$

So, by the uniqueness of the spherical function with eigenvalue λ we deduce that $F_x(y) = C_x\varphi^\lambda(y)$ for some constant C_x. To compute $C_x = F_x(1)$ we take: $F_x(1) = \int_K f(xk)dk = \int_K f(x)dk = f(x)$ and the lemma is proven. □

We can now continue the proof of the theorem: To compute $\varphi(g) = \langle\rho(g)f, f\rangle = \langle\rho(k)\rho(g)f, f\rangle$ we write:

$$\langle\rho(g)f, f\rangle = \int_{\Gamma\backslash G} f(xg)\overline{f(x)}dx = \int_K\int_{\Gamma\backslash G} f(xkg)\overline{f(kx)}dkdx$$
$$= \int_{\Gamma\backslash G}\int_K f(xkg)dk\overline{f(x)}dx = \int_{\Gamma\backslash G} \varphi^\lambda(g)f(x)\overline{f(x)}dx$$
$$= \varphi^\lambda(g)\|f\|^2 = \varphi^\lambda(g).$$

(The first equality is just the definition, the second is obtained by replacing x by xk and integrating over K, for the third recall that f is right K-invariant, and the fourth follows from Lemma 5.5.4.) So we have proved that $\varphi = \varphi^\lambda$, which proves that ρ^λ occurs in $L^2(\Gamma\backslash G)$.

This side of the proof works almost word by word also for the case of G_∞.

Now assume ρ^λ occurs in $L^2(\Gamma\backslash G)$. Again we begin with $G = G_p$.

Since ρ^λ occurs in $L^2(\Gamma\backslash G)$ there is a K-invariant vector f, in the subspace on which G acts via ρ^λ. We want to prove that $\delta(f) = \lambda f$.

Since $\delta(f) = f * \overline{\delta}$, i.e., convolution with a characteristic function of a subset of G (more specifically with the characteristic function of $K\begin{pmatrix} p & 0 \\ 0 & 1\end{pmatrix}K$), $\delta(f)$ is at the same submodule of $L^2(\Gamma\backslash G)$ as f. Moreover, it is also right-K-invariant and hence by (5.1.4), $\delta(f) = cf$ for some c. To prove that $c = \lambda$ we use the following general simple Lemma which is of independent interest.

Lemma 5.5.5. *Let (H, ρ) be an irreducible unitary representation of G of class one with associated spherical function φ. Let v be a K-invariant vector. Then for every $g \in G$, $\int_K \rho(kg)vdk = \varphi(g)v$.*

Proof. $\int_K \rho(kg)v dk$ is K-invariant and hence a multiple of v. To compute that scalar, let:

$$\begin{aligned}\langle \int_K \rho(kg)v dk, v\rangle &= \int_K \langle \rho(g)v, \rho(k^{-1})v\rangle dk \\ &= \int_K \langle \rho(g)v, v\rangle dk = \varphi(g).\end{aligned}$$

The lemma is, therefore, proven. □

For our case this implies that for every x and g in G,

$$\varphi^\lambda(g)f(x) = \int_K \rho(kg)f(x)dk = \int_K f(xkg)dk.$$

Now, in order to compute the constant c, let $g \in G$ with $f(g) \neq 0$, and assume that $f(g) = 1$. We remember that f is defined on the tree G/K and as such $\delta(f)(g)$ is just the sum over the neighbors of g. So,

$$\begin{aligned}\delta(f)(g) &= \sum_{y \sim g} f(y) = (p+1)\int_K f\left(gk\begin{pmatrix} p & 0 \\ 0 & 1\end{pmatrix}\right)dk \\ &= (p+1)\varphi^\lambda\begin{pmatrix} p & 0 \\ 0 & 1\end{pmatrix} f(g) = (p+1)\frac{\lambda}{p+1} = \lambda.\end{aligned}$$

The third equality follows from Lemma 5.5.5 and the fourth one from the computation we made in Section 5.3 which shows that $\varphi^\lambda\begin{pmatrix} p & 0 \\ 0 & 1\end{pmatrix} = \frac{\lambda}{p+1}$. This completes the proof of Theorem 5.5.1 for the case $G = G_p$.

The proof for $G = G_\infty$ is similar, but some remarks are in order. We start with the following lemma which gives a useful expression for the Laplacian in this case and is therefore of independent interest:

Lemma 5.5.6. *Let h be a C^∞-function on $\mathbb{H}$, and $g_t = \begin{pmatrix} 1 & t \\ 0 & 1\end{pmatrix}$ for $t \in [0,1]$. Then for every $g \in G$,*

$$\Delta(h)(g) = -4 \cdot \lim_{t\to 0} \frac{\int_K h(gkg_t)dk - h(g)}{t^2}.$$

Proof. Both expressions are «local». Moreover, it suffices to prove it in case h is an eigenfunction of the Laplacian Δ, say $\Delta(h) = \lambda h$ for some λ.

A computation identical to Lemma 5.5.4 shows that for every $x, y \in G$, $\int_K h(xky)dk = \varphi^\lambda(y)h(x)$. Hence

$$-4 \cdot \lim_{t\to 0} \frac{\int_K h(gkg_t)dk - h(g)}{t^2} = -4 \cdot \lim_{t\to 0}\left(\frac{\varphi^\lambda(g_t) - 1}{t^2}\right)h(g).$$

But now as $\varphi^\lambda(i) = 1$ and $(\varphi^\lambda)'(i) = 0$ (since i corresponds to the identity coset of G_∞/K), the expression we obtain is $(-4) \cdot \frac{(\varphi^\lambda)''(i)}{2} h(g) = -2(\varphi^\lambda)''(i)h(g)$. We know that $\Delta(\varphi^\lambda) = \lambda\varphi^\lambda$ and also that (since φ^λ is spherical) $\frac{\partial^2}{\partial x^2}\varphi^\lambda\Big|_i = \frac{\partial^2}{\partial y^2}\varphi^\lambda\Big|_i = -\frac{1}{2}\lambda$. Hence $-2(\varphi^\lambda)''(i) = (-2)\left(\frac{-\lambda}{2}\right) = \lambda$ and the lemma is proved.

One can now argue for the proof of Theorem 5.5.1 for G_∞ just as for G_p. The operator Δ was expressed as a limit of averaging operators. Thus if $f \in L^2(\Gamma\backslash G)$ is K-invariant (and gives rise to the spherical function φ^λ), $\Delta(f)$ lies in the same submodule as f. Since f is essentially unique in the irreducible subrepresentation it generates, and $\Delta(f)$ is also K-invariant, we conclude that $\Delta(f) = cf$ for some constant c.

To compute c: By Lemma 5.5.5, $\int_K f(xkg)dk = \varphi^\lambda(g)f(x)$ and so

$$\begin{aligned}
\Delta(f)(g) &= -4 \cdot \lim_{t\to 0} \frac{\int_K f(gkg_t)dk - f(g)}{t^2} \\
&= -4 \cdot \lim_{t\to 0} \frac{\varphi^\lambda(g_t) - 1}{t^2} f(g) = (-4)\frac{(\varphi^\lambda)''(i)}{2} f(g) \\
&= (-4)\frac{-\lambda}{2\cdot 2} f(g) = \lambda f(g),
\end{aligned}$$

and the theorem is proven. □

6 Spectral Decomposition of $L^2(G(\mathbb{Q})\backslash G(\mathbb{A}))$

6.0 Introduction

This chapter begins with a presentation of the adèles of the rational numbers. This locally compact ring $\mathbb{A} = \prod' \mathbb{Q}_p$ is a restricted product of all possible completions of $\mathbb{Q}$ – the field of rational numbers. When G is a real or p-adic algebraic group and Γ a congruence subgroup Strong Approximation results enable us to embed $L^2(\Gamma\backslash G)$ into $L^2(G(\mathbb{Q})\backslash G(\mathbb{A}))$ (see Section 6.3 for a precise formulation). The adèles in general and the spectral decomposition of $L^2(G(\mathbb{Q})\backslash G(A))$ in particular give a convenient way to state results on «all» spaces $L^2(\Gamma\backslash G)$ together. This way we will see, for example, that the Selberg conjecture is a special case of a more general conjecture which asserts that no complementary series representations appear as local factors of subrepresentation of $L^2(PGL(\mathbb{Q})\backslash PGL_2(\mathbb{A}))$. Another special case of this general conjecture is a theorem of Deligne (see (6.2.2)). This theorem, which is, in fact, a representation theoretic reformulation of Ramanujan Conjecture (known also as Petersson's conjecture), is extremely important to us. This is the theorem which is responsible for both the final solution of the Banach-Ruziewicz problem and the construction of Ramanujan graphs. (To be precise we should say that both problems can be solved using some weaker results proved earlier by Rankin and Eichler, respectively, but Deligne's Theorem gives a unified approach as well as better results in the Ruziewicz problem (see Chapter 9).)

The solution of these problems will be presented in Chapter 7. For it we will not need the spectral decomposition of $L^2(PGL_2(\mathbb{Q})\backslash PGL(\mathbb{A}))$ but rather of $L^2(G'(\mathbb{Q})\backslash G'(\mathbb{A}))$ where G' is the group of invertible elements of the Hamiltonian quaternion algebra modulo its center (or any other definite quaternion algebra). Due to Jacquet-Langlands correspondence (6.2.1), Deligne's Theorem gives the desired information also about $G'(\mathbb{A})$.

In this chapter we bring merely a survey. A more detailed exposition from a broader prespective and with more proofs (or at least hints on proofs) is given at the end of the book in the Appendix, written by Jon Rogawski.

6.1 Deligne's Theorem; adèlic formulation

In this section we introduce the language of adèles as a convenient way to state the theorem of Deligne (which is actually the solution to the classical Ramanujan conjecture – see the Appendix).

On the field $\mathbb{Q}$ of rational numbers there are two types of valuations: archimedian and non-archimedian. There is, essentially, a unique archimedian valuation – the standard absolute value $v_\infty(a) = |a|$. For each rational prime p, there is a p-adic valuation defined as follows: For $a \in \mathbb{Q}$, set $a = p^{v_p(a)}b$, where $b = \frac{m}{n}$ both

m and n are integers prime to p and $v_p(a) \in \mathbb{Z}$. Then define $|a|_p = p^{-v_p(a)}$. It is well known (see [BSh]) that the above examples cover all valuations of $\mathbb{Q}$. A valuation defines a metric (by $d(a,b) = |a-b|$) and the completions of $\mathbb{Q}$ with respect to these valuations are respectively $\mathbb{Q}_\infty = \mathbb{R}$ and $\mathbb{Q}_p$ – the field of p-adic numbers. We regard ∞ as the infinite prime, and let p run over all primes – the finite ones and the infinite one.

For a *finite* prime p, we have in $\mathbb{Q}_p$ a compact open subring $\mathbb{Z}_p$ – the ring of p-adic integers which is characterized as $\mathbb{Z}_p = \{x \in \mathbb{Q}_p \mid |x|_p \leq 1\}$.

Definition 6.1.1. *Let*

$$\mathbb{A} = \{\alpha = (\dots, a_p, \dots) \in \prod_p \mathbb{Q}_p | a_p \in \mathbb{Z}_p \text{ for almost all } p\}$$

where «almost all» means «for all but a finite set», and the product runs over all (finite and infinite) primes $p = \infty, 2, 3, 5, \dots$. $\mathbb{A}$ is a ring with the componentwise addition and multiplication. We put a topology on $\mathbb{A}$ by declaring $\mathbb{R} \times \hat{\mathbb{Z}} = \mathbb{R} \times \prod_{p \text{finite}} \mathbb{Z}_p$, with its Tychonof product topology, to be an open subring of $\mathbb{A}$. $\mathbb{A}$ is a locally compact ring called the Adèle ring of $\mathbb{Q}$.

The field $\mathbb{Q}$ can be embedded diagonally in $\mathbb{A}$ by $a \mapsto (a, a, a, \dots)$ for $a \in \mathbb{Q}$. It is not difficult to check that $\mathbb{Q}$ is a discrete subring and that $\mathbb{Q} + ([0,1] \times \hat{\mathbb{Z}}) = \mathbb{A}$ and so $\mathbb{Q}$ is a cocompact lattice in $\mathbb{A}$. (For more details on $\mathbb{A}$, see [GGPS].)

Let $GL_2(\mathbb{A})$ be the group of invertible 2×2 matrices over $\mathbb{A}$. It can also be defined, as in (6.1.1), as a restricted direct product, i.e.,

$$GL_2(\mathbb{A}) \simeq \{g = (g_p) \in \prod_p GL_2(\mathbb{Q}_p) | g_p \in GL_2(\mathbb{Z}_p) \text{ for almost all } p\}.$$

Similarly, for $SL_2(\mathbb{A})$, $PGL_2(\mathbb{A}) = GL_2(\mathbb{A})/Z$ where Z is its center, etc. Now, since $\mathbb{Q}$ is a cocompact discrete subring of $\mathbb{A}$, it behaves much as $\mathbb{Z}$ inside $\mathbb{R}$. It is therefore not surprising that for many algebraic groups G defined over $\mathbb{Q}$, $G(\mathbb{Q})$ is a lattice in $G(\mathbb{A})$. So for example, $SL_2(\mathbb{Q})$ (resp: $PGL_2(\mathbb{Q})$) is a lattice in $SL_2(\mathbb{A})$ (resp: $PGL_2(\mathbb{A})$). (But $GL_2(\mathbb{Q})$ is not of finite covolume in $GL_2(\mathbb{A})$, just like $GL_2(\mathbb{Z})$ is not in $GL_2(\mathbb{R})$). For the general result in this direction, see [MT].

Unitary irreducible representations ρ of $G(\mathbb{A})$ (say G is SL_2 or PGL_2) are always «completed» tensor products $\rho = \otimes \rho_p$ where ρ_p is an irreducible unitary representation of $G(\mathbb{Q}_p)$ (and ρ_p is of class one for almost all p) (see [GGPS, p. 274]). ρ_p is called the p-factor (or the local p-factor) of ρ.

We are now in a position to state the following theorem of Deligne. In the Appendix one can see it in its more familiar form as a theorem estimating Fourier coefficients of modular cusp forms and the equivalence between the two forms is explained. For our applications here the following representation theoretic formulation is more suitable.

Theorem 6.1.2. *Let $G = PGL_2(\mathbb{A})$, $\Gamma = PGL_2(\mathbb{Q})$ and $\rho = \otimes\rho_p$ an irreducible (not one-dimensional) subrepresentation of G in $L^2(\Gamma\backslash G)$. Assume ρ_∞ is a discrete series representation of $PGL_2(\mathbb{R})$ (i.e., ρ_∞ is a subrepresentation of $L^2(PGL_2(\mathbb{R}))$). Then for every finite p, ρ_p is not in the complementary series of representations of $PGL_2(\mathbb{Q}_p)$.*

In [Vi2] M.F. Vigneras mentions a stronger conjecture:

Conjecture 6.1.3. *Let G, Γ, ρ be as in 6.1.2 – but without any assumption on ρ_∞. Then for every p (finite or infinite) ρ_p is not in the complementary series.*

In Chapter 7 we will see how powerful Theorem 6.1.2 is. But to relate it to classical arithmetic problems and to our (not so classical...) graph and measure theoretic problems we need to relate it back to $L^2(\Gamma\backslash G)$ where G is a real or p-adic Lie group and Γ is a lattice in G. The Strong Approximation Theorem is what enables us to do this.

6.2 Quaternion algebras and groups

In this section we bring the Jacquet-Langlands correspondence which enables us to extend Theorem 6.1.2 to a wider class of groups. More specifically to groups of units of quaternion algebras. This machinery provides a method to transfer results on spectral decomposition of non-uniform lattices in $PSL_2(\mathbb{R})$ to uniform lattices as we did in Section 4.4. But even more important for us is the fact that the correspondence will enable us to deduce results on cocompact lattices in $PGL_2(\mathbb{Q}_p)$. We begin with some background on quaternion algebras.

Let u and v be positive rational number, $D = D(u,v)$ a four-dimensional algebra over $\mathbb{Q}$ with the basis $1, i, j$ and k subject to the relations

$$i^2 = -u, j^2 = -v \text{ and } k = ij = -ji.$$

So an element $\alpha \in D$ is of the form $\alpha = x_0 + x_1 i + x_2 j + x_3 k$ whose *conjugate* is $\overline{\alpha} = x_0 - x_1 i - x_2 j - x_3 k$ and whose *norm* $N(\alpha)$ and *trace* $\mathrm{Tr}(\alpha)$ are defined to be $N(\alpha) = \alpha\overline{\alpha} = x_0^2 + ux_1^2 + vx_2^2 + uvx_3^2$ and $\mathrm{Tr}(\alpha) = \alpha + \overline{\alpha} = 2x_0$. The norm is multiplicative $N(\alpha\beta) = N(\alpha)N(\beta)$. If $\mathbb{F}$ is a field containing $\mathbb{Q}$ (e.g., $\mathbb{R}$ or $\mathbb{Q}_p$) then $D_\mathbb{F} = \mathbb{F} \otimes_\mathbb{Q} D$ is an $\mathbb{F}$-algebra. There are two possibilities: either $D_\mathbb{F}$ is a division algebra (i.e., every nonzero element α of $D_\mathbb{F}$ has an inverse: this happens iff $N(\alpha) \neq 0$ for every $\alpha \neq 0$ in which case $\alpha^{-1} = \frac{\overline{\alpha}}{N(\alpha)}$) or $D_\mathbb{F} \simeq M_2(\mathbb{F})$, the algebra of 2×2 matrices over $\mathbb{F}$. In the first case we say that D *ramifies* or *does not split* in $\mathbb{F}$, and that it *splits* or *unramified* in the second case. In that second case there is an isomorphism $\sigma : D_\mathbb{F} \to M_2(\mathbb{F})$ satisfying $\det(\sigma(\alpha)) = N(\alpha)$ and $\mathrm{Tr}\,\sigma(\alpha) = \mathrm{Tr}(\alpha)$ for every $\alpha \in D_\mathbb{F}$.

It is well known (cf. [Vi1, Chapter II]) that $D(u,v)$ splits in $\mathbb{Q}_p$ ($p \leq \infty$) if and only if the equation $ux^2 + vy^2 - z^2 = 0$ has a non-trivial solution in $\mathbb{Q}_p$.

This happens for almost all p; the exceptional set of primes is of even cardinality. For example, the standard Hamiltonian quaternion algebra $D = D(1,1)$ ramifies only at $p = 2, \infty$ and splits at all the others. It is especially easy to present an isomorphism $D_{\mathbb{Q}_p} \simeq M_2(\mathbb{Q}_p)$ in case $p \equiv 1 \pmod 4$, where there is in $\mathbb{Q}_p$ a square root of -1, say $\varepsilon^2 = -1$. In this case $\sigma : D_{\mathbb{Q}_p} \to M_2(\mathbb{Q}_p)$ is given by:

$$\sigma(x_0 + x_1 i + x_2 j + x_3 k) = \begin{pmatrix} x_0 + x_1\varepsilon & x_3 + x_2\varepsilon \\ -x_3 + x_2\varepsilon & x_0 - x_1\varepsilon \end{pmatrix}.$$

This is an isomorphism as can be checked directly. More generally, if $-u$ and $-v$ are squares in $\mathbb{F}$, then the isomorphism between $D_{\mathbb{F}}$ and $M_2(\mathbb{F})$ is given by:

$$\sigma(x_0 + x_1 i + x_2 j + x_3 k) = \begin{pmatrix} x_0 + \sqrt{-u}x_1 & \sqrt{-v}(x_2 + \sqrt{-u}x_3) \\ \sqrt{-v}(x_2 - \sqrt{-u}x_3) & x_0 - \sqrt{-u}x_1 \end{pmatrix}.$$

It is also known that the $\mathbb{Q}$-quaternion algebra D is determined by the finite set S of (finite and infinite) primes for which D is ramified in $\mathbb{Q}_p$ iff $p \in S$ (cf. [Re]). The algebra D is, therefore, sometimes denoted D_S, e.g., the Hamiltonian quaternions $D(1,1)$ can also be denoted $D_{\{2,\infty\}}$.

Let $G'(\mathbb{F})$ be the group of invertible elements of $D_{\mathbb{F}}$ modulo its center. So if D splits over $\mathbb{F}$ then $G'(\mathbb{F}) \simeq PGL_2(\mathbb{F})$. For the Hamiltonian quaternions $G'(\mathbb{R}) \simeq SU(2)/\{\pm 1\} \simeq SO(3)$. In fact, for those primes p for which D ramifies at Q_p, $G'(\mathbb{Q}_p)$ is compact. As in Section 6.1, $G'(\mathbb{Q})$ is a lattice in $G'(\mathbb{A})$.

We have by now all the notation to state the following theorem of Jacquet and Langlands:

Theorem 6.2.1. ([JL] see [Ge, Theorems 10.1 and 10.2]) *Let $D = D_S$ be a quaternion algebra defined over $\mathbb{Q}$ and ramified at the set of primes S. Let G' be the algebraic group defined by $G'(\mathbb{F}) = (\mathbb{F} \otimes_{\mathbb{Q}} D)^*/Z(\mathbb{F} \otimes_{\mathbb{Q}} D)^*$ where $(:)^*$ denotes the invertible elements and $Z(:)$ the center.*

Let $\rho' = \otimes_p \rho'_p$ be an irreducible, not one-dimensional $G'(\mathbb{A})$ subrepresentation of $L^2(G'(\mathbb{Q})\backslash G'(\mathbb{A}))$. Then there exists an irreducible subrepresentation $\rho = \otimes \rho_p$ of $L^2(PGL_2(\mathbb{Q})\backslash PGL_2(\mathbb{A}))$ such that:

1. *if $p \notin S$, $\rho_p = \rho'_p$.*
2. *if $p \in S$, then ρ_p is a discrete series representation, i.e., isomorphic to a subrepresentation of $PGL_2(\mathbb{Q}_p)$ acting on $L^2(PGL_2(\mathbb{Q}_p))$.*

Note that if $p \notin S$ then D splits over $\mathbb{Q}_p$ so $G'(\mathbb{Q}_p) = PGL_2(\mathbb{Q}_p)$, and hence it is possible to write ρ'_p in (i).

Theorem 6.2.1 combined with (6.1.2) gives the following important corollary:

Corollary 6.2.2. *Let D be a definite quaternion algebra (i.e., an algebra which ramifies at $\mathbb{Q}_\infty = \mathbb{R}$, e.g., the standard Hamiltonian algebra). Let $\rho' = \otimes_p \rho'_p$ be an irreducible, not one-dimensional, subrepresentation of $L^2(G(\mathbb{Q})\backslash\, G'(\mathbb{A}))$. Then for every prime p for which D splits (and so ρ'_p is a representation of $PGL_2(\mathbb{Q}_p)$), ρ'_p is not in the complementary series of representations of $PGL_2(\mathbb{Q}_p)$.*

6.3 The Strong Approximation Theorem and its applications

In this section we apply the Strong Approximation Theorem (in the special cases relevant to us) to present the connection between $G(\mathbb{Q})\backslash G(\mathbb{A})$ and $\Gamma(m\mathbb{Z})\backslash G(\mathbb{R})$ or various other homogeneous spaces. This will enable us to apply the previous results of this chapter. Most of the applications will be given in the next chapter. Here we merely present Selberg's Conjecture as a special case of Conjecture 6.1.3 (complementary to Deligne's Theorem 6.1.2).

The main technical tool here is the Strong Approximation Theorem, which asserts that for a $\mathbb{Q}$-algebraic group G (under suitable assumptions – simply connected and absolutely almost simple – see [Kn] and [Pr2]), $G(\mathbb{Q})G(\mathbb{Q}_p)$ is dense in $G(\mathbb{A})$ for every prime p for which $G(\mathbb{Q}_p)$ is not compact. Assume G satisfies these hypotheses; then for every open subgroup K of $G(\mathbb{A})$ we have: $G(\mathbb{Q})G(\mathbb{Q}_p)K = G(\mathbb{A})$. For example, take $p = \infty$ and $K = G(\mathbb{R}) \times \prod_{p<\infty} K'_p$ where K'_p is an open (finite index) subgroup of $K_p = G(\mathbb{Z}_p)$ and $K'_p = K_p$ for almost all p. More specifically, for $0 < N \in \mathbb{Z}$ with prime decomposition $N = \prod_{i=1}^{r} p_i^{\alpha_i}$, take:

$$K'_{p_i} = \ker(G(\mathbb{Z}_p) \to G(\mathbb{Z}_p/p_i^{\alpha_i}\mathbb{Z}_p)) \quad \text{for} \quad i = 1, \ldots, r,$$

$$K'_p = K_p \quad \text{for} \quad p \neq p_i, \quad K^N = G(\mathbb{R}) \times \prod_{p<\infty} K'_p \quad \text{and} \quad K_0^N = \prod_{p<\infty} K'_p.$$

We then have:

$$G(\mathbb{A}) = G(\mathbb{Q})G(\mathbb{R})K_0^N.$$

This yields:

$$G(\mathbb{Q})\backslash G(\mathbb{A})/K_0^N = \Gamma(N)\backslash G(\mathbb{R}) \tag{$*$}$$

where $\Gamma(N)$ is the N-congruence subgroup of $\Gamma = G(\mathbb{Z})$, i.e., $\Gamma(N) = \ker(G(\mathbb{Z}) \to G(\mathbb{Z}/N\mathbb{Z}))$. (Note that $\Gamma(N) = G(\mathbb{Q}) \cap K^N$ which is the reason why $\Gamma(N)$ comes up in formula $(*)$. See [GGPS, pp. 352–356] for more details.) Note also that in the left-hand side of $(*)$ we divided $G(\mathbb{Q})\backslash G(\mathbb{A})$ by K_0^N and not by K_N. This leaves room for the action of $G(\mathbb{R})$, which is isomorphic to the $G(\mathbb{R})$ action on the right-hand side. So we also have the following isomorphism of $G(\mathbb{R})$-modules:

$$L^2(G(\mathbb{Q})\backslash G(\mathbb{A})/K_0^N) \simeq L^2(\Gamma(N)\backslash G(\mathbb{R})).$$

This shows that every $G(\mathbb{R})$-submodule of $L^2(\Gamma(N)\backslash G(\mathbb{R}))$ is in fact a $G(\mathbb{R})$-submodule of $L^2(G(\mathbb{Q})\backslash G(\mathbb{A}))$.

Now the time comes to be more specialized. The assumption of the Strong Approximation Theorem applies to SL_2, but it is not quite true for PGL_2. Still, from the Strong Approximation Theorem for SL_2 one can deduce the following for PGL_2.

Proposition 6.3.1. *In the notation above: $H = PGL_2(\mathbb{Q}) \cdot PGL_2(\mathbb{R}) \cdot K_0^N$ is a finite index normal open subgroup of $G = PGL_2(\mathbb{A})$. The map*

$$PGL_2(\mathbb{Q})\backslash PGL_2(\mathbb{A})/K_0^N \to G/H \qquad (**)$$

decomposes the left-hand side of $(**)$ into finitely many fibers. The principal one (i.e., the pre-image of $1H$) is canonically isomorphic to $\Gamma(N)\backslash PGL_2(\mathbb{R})$, where

$$\Gamma(N) = \ker(PGL_2(\mathbb{Z}) \to PGL_2(\mathbb{Z}/N\mathbb{Z})).$$

Moreover, $L^2(\Gamma(N)\backslash PGL_2(\mathbb{R}))$ is isomorphic as an $PGL_2(\mathbb{R})$-module to a submodule of $L^2(PGL_2(\mathbb{Q})\backslash PGL_2(\mathbb{A})/K_0^N)$.

Proof. Everything follows from the fact that H is a finite index normal subgroup of G. That fact follows from the Strong Approximation Theorem for SL_2 together with the decomposition of $\mathbb{A}^*$ (= the invertible elements of $\mathbb{A}$ = the *idele* group of $\mathbb{Q}$) as $\mathbb{A}^* = \mathbb{Q}^* \cdot \mathbb{R}^* \cdot \prod_{p<\infty} \mathbb{Z}_p^*$. The details are left to the reader. □

Proposition 6.3.2. *Conjecture 6.1.3 implies Selberg's conjecture $\lambda_1 \geq \frac{1}{4}$ (see (4.4.1)).*

Proof. From (6.1.3) it would follow, using (6.3.1), that the complementary series representations of $PGL_2(\mathbb{R})$ do not appear as subrepresentations of $L^2(\Gamma(N)\backslash PGL_2(\mathbb{R}))$. From this, it follows that the complementary series representations of $PSL_2(\mathbb{R})$ do not appear in $L^2(\Gamma(N)\backslash PSL_2(\mathbb{R}))$. The last assertion is equivalent to Selberg's conjecture by (5.5.2). □

It is now clear how Deligne's Theorem 6.1.2 is related to Selberg's conjecture: They do not imply each other, but both are special cases of Conjecture 6.1.3. Note that Conjecture 6.1.3 is still stronger that the union of the two!

We are now going to take another special case of the Strong Approximation Theorem, this time for quaternion groups similar to those discussed in Section 6.2.

Let D be a *definite* quaternion algebra defined over $\mathbb{Q}$ (i.e., D does not split over $\mathbb{R}$ and so $\mathbb{R} \otimes_{\mathbb{Q}} D$ is a division algebra isomorphic to the real Hamiltonian quaternions). Let G' be the $\mathbb{Q}$-algebraic group of the invertible elements of D modulo its center. So $G'(\mathbb{R}) \simeq SO(3)$. Let p be a finite prime for which $\mathbb{Q}_p \otimes_{\mathbb{Q}} D$ splits (and so $G(\mathbb{Q}_p) \simeq PGL_2(\mathbb{Q}_p)$). By (3.2.2D), $\Gamma = G'\left(\mathbb{Z}\left[\frac{1}{p}\right]\right)$ is a lattice in $G'(\mathbb{R}) \times G'(\mathbb{Q}_p)$. Let's denote $\Gamma(N) = \ker(G'\left(\mathbb{Z}\left[\frac{1}{p}\right]\right) \to G'\left(\mathbb{Z}\left[\frac{1}{p}\right]/N\mathbb{Z}\left[\frac{1}{p}\right]\right))$ for $(p, N) = 1$. Also, as before, $K_0^N = \prod_{p\neq\ell<\infty} K'_\ell$ where $K'_\ell = G'(\mathbb{Z}_\ell)$ for $\ell \nmid N$ and $\ell \neq p$, and $K'_{\ell_i} = \ker(G'(\mathbb{Z}_{\ell_i}) \to G'(\mathbb{Z}_{\ell_i}/\ell_i^{d_i}\mathbb{Z}_{\ell_i}))$ when $N = \prod_{i=1}^r \ell_i^{d_i}$.

An argument similar to (6.3.1) (see also [Vi1, p. 81]) gives:

Proposition 6.3.3. *$H = G'(\mathbb{Q}) \cdot G'(\mathbb{R}) \cdot G'(\mathbb{Q}_p) \cdot K_0^N$ is a finite index normal subgroup of $G'(\mathbb{A})$. Also $G'(\mathbb{Q})\backslash H/K_0^N \simeq \Gamma(N)\backslash G'(\mathbb{R}) \times G'(\mathbb{Q}_p)$, and $L^2(\Gamma(N)\backslash G'(\mathbb{R}) \cdot G'(\mathbb{Q}_p))$ is isomorphic as a $G'(\mathbb{R}) \times G'(\mathbb{Q}_p)$-module to a subrepresentation of $L^2(PGL_2(\mathbb{Q})\backslash G'(\mathbb{A})/K_0^N)$.*

To illustrate the way the results of this chapter are going to be applied in the next chapter, we sketch here a proof of the following, which has been already used in Theorem (4.4.3) without a proof.

Theorem 6.3.4. *Let D be an indefinite quaternion algebra such that $D_{\mathbb{Q}}$ is a division algebra but $D_{\mathbb{R}} \simeq M_2(\mathbb{R})$ (see [GGPS, p. 116] for examples). Let G' be as before. Then:*

1. *$\Gamma = G'(\mathbb{Z})$ is a cocompact lattice in $G'(\mathbb{R}) = PGL_2(\mathbb{R})$.*
2. *For $\Gamma(N) = \ker(G'(\mathbb{Z}) \to G'(\mathbb{Z}/N\mathbb{Z}))$, we have $\lambda_1(\Gamma(N)\backslash\mathbb{H}) \geq \frac{3}{16}$.*

Proof.

1. This is a special case of (3.2.1). The fact that it is cocompact follows from a general criterion ([Gd], [MT]) which applies here since $D_{\mathbb{Q}}$ is a division algebra. For a direct proof see [GGPS, p. 117].
2. We already know from (5.5.1) that a question on eigenvalues λ is indeed one on the irreducible subrepresentations of $L^2(\Gamma(N)\backslash PGL_2(\mathbb{R}))$. Such an irreducible representation ρ can be «extended», using the Strong Approximation Theorem (as in (6.3.3)), to an irreducible subrepresentation ρ' of $L^2(G'(\mathbb{Q})\backslash G'(\mathbb{A}))$. Now we have the Jacquet-Langlands correspondence (6.2.1). This corresponds to ρ' an irreducible subrepresentation $\widetilde{\rho} = \otimes_p \widetilde{\rho}_p$ of $L^2(PGL_2(\mathbb{Q})\backslash PGL_2(\mathbb{A}))$ with $\widetilde{\rho}_\infty = \rho$. The infinite factor $\widetilde{\rho}_\infty$ of such a representation is governed by Selberg's Theorem $\lambda_1 \geq \frac{3}{16}$, which means that if $\widetilde{\rho}_\infty = \rho_s$, then either $s \in i\mathbb{R}$ or $s \in \mathbb{R}$ and $|s| \leq \frac{1}{2}$ (see (5.2.3)). So the same representation theoretic restriction applies to $L^2(\Gamma(N)\backslash PGL_2(\mathbb{R}))$; hence $\lambda_1 \geq \frac{3}{16}$ for these congruence subgroups as well. □

6.4 Notes

1. The general spirit of this chapter follows M.F. Vigneras' paper [Vi2]. The details follow Gelbart [Gel] and Gelfand-Graev-Piatetski-Shapiro [GGPS]. Both books contain much more information and details in this direction. In particular, Gelbart's book contains a proof of the Jacquet-Langlands correspondence (6.2.1).
2. Deligne's Theorem (6.1.2) as explained in detail in the Appendix is an equivalent form of the classical Ramanujan conjecture (or Peterson-Ramanujan conjecture as it is called in some books, or Petersson's conjecture in others). It was proved by Deligne in two steps. At first ([Dl1]) it was shown to follow

from Weil's conjectures and then in [Dl2] Weil's conjectures were proved. Before that some special cases were proved by Eichler [Ei] by a reduction to Weil's Theorem («Riemann hypothesis for curves over finite fields»).

3. It would be more precise in the statement of (6.1.2) (and some of the other results in this chapter) to use the term «cuspidal representation». We avoided making up this additional assumption (i) by talking about subrepresentations – so by this we mean, in particular, representations appearing in the discrete spectrum – and (ii) by assuming the representation is not one-dimensional. The reader should be aware that in more general groups the discrete spectrum is not only «cuspidal». Fortunately, we don't have to deal with it. But the generalization of the Peterson-Ramanujan Conjecture to other groups is not quite the same. (In any way it was shown that the naive generalizations are not true, see [BuS] and the references therein.)
4. M.F. Vigneras' book [Vi1] is a good source of information about quaternion algebras, their groups and the connection between the classical and the adelic formulation.

7 Banach-Ruziewicz Problem for $n = 2, 3$; Ramanujan Graphs

7.0 Introduction

In this chapter we (finally...) pick the fruits: We will prove Drinfeld's Theorem that the Lebesgue measure is the only finitely additive rotationally invariant measure defined on the Lebesgue measurable subsets of the sphere S^n for $n = 2, 3$. (For $n \geq 4$, this was proved in Chapter 3, while for $n = 1$, it is not true! See (2.2.11).) We will also construct, for every prime p, a family of $(p+1)$-regular Ramanujan graphs (in particular, giving rise to the best known expanders!).

Both problems will be solved using the same group Γ ! Let's fix now the notations for the rest of the chapter: Let $D = D(u, v)$ be a definite quaternion algebra defined over $\mathbb{Q}$, (i.e., $u, v > 0$, for example, the standard Hamiltonian quaternion algebra), G' is the $\mathbb{Q}$-algebraic group of the invertible elements of D modulo the central ones, i.e., $G' = D^*/Z(D^*)$. Let p be a finite prime for which D splits in $\mathbb{Q}_p$ (e.g., p is any odd prime in the case of the Hamiltonian quaternions) and $\Gamma = G'(\mathbb{Z}[\frac{1}{p}])$. Then the diagonal embedding $\Gamma \to G'(\mathbb{R}) \times G'(\mathbb{Q}_p) \simeq SO(3) \times PGL_2(\mathbb{Q}_p)$ presents Γ as a cocompact lattice in the «mixed» Lie group $G = SO(3) \times PGL_2(\mathbb{Q}_p)$. If N is a positive integer prime to p, say $N = \prod_{i=1}^{r} \ell_i^{\alpha_i}$, we denote $\Gamma(N) = \ker(G'(\mathbb{Z}[\frac{1}{p}]) \to G'(\mathbb{Z}[\frac{1}{p}]/N\mathbb{Z}[\frac{1}{p}])$ and $\Gamma(N)$ is called the N-congruence subgroup of Γ. Being a finite index subgroup of Γ, it is also a cocompact lattice of G. The main result, from which the two solutions follow, is the following:

Theorem 7.1.1. *If $\rho = \rho_\infty \otimes \rho_p$ is an irreducible subrepresentation of $L^2(\Gamma(N)\backslash G)$ then ρ_p is not from the complementary series (unless it is one-dimensional).*

This theorem is proved in 7.1 using all the machinery we prepared in Chapter 6. Then by looking at the projection of Γ into the real factor of G, we will prove Drinfeld's Theorem which is the affirmative solution to the Banach-Ruziewicz problem for $n = 2, 3$. The method is essentially the method of Drinfeld but we use a different group (Γ instead of $G'(\mathbb{Q})$). This has the advantage of being «effective». (See more in Chapter 9.) It was shown by Sarnak [Sa2] that this affirmative solution for $n = 2$, in fact, implies an affirmative solution for every $n \geq 2$ (and even an effective one), but we will not go into this here as we already proved it for $n \geq 4$ (though, in a «non-effective» way) and the case $n = 3$ can be deduced easily with the case $n = 2$.

In 7.3 we use the same group Γ (or more precisely, its congruence subgroups $\Gamma(N)$) to construct Ramanujan graphs. This time we look at the projection of $\Gamma(N)$ into the p-adic factor $PGL_2(\mathbb{Q}_p)$. The graphs are $\Gamma(N)\backslash PGL_2(\mathbb{Q}_p)/PGL_2(\mathbb{Z}_p)$. These are finite graphs since $PGL_2(\mathbb{Q}_p)/PGL_2(\mathbb{Z}_p)$ is a tree and $\Gamma(N)$ is cocompact in $PGL_2(\mathbb{Q}_p)$. The construction here follows Lubotzky-Phillips-Sarnak

[LPS1, LPS2] and Margulis [M6]. We shall also estimate some more combinatorial invariants of these graphs – such as girth, chromatic numbers and diameter. Finally, in 7.4, we will give a «more explicit» presentation of them in some special cases.

7.1 The spectral decomposition of $G'(\mathbb{Z}[\frac{1}{p}])\backslash G'(\mathbb{R}) \times G'(\mathbb{Q}_p)$

We continue here with the notations set up at the introduction of this chapter.

Theorem 7.1.1. *Let $\rho = \rho_\infty \otimes \rho_p$ be a unitary irreducible representation of $G = G'(\mathbb{R}) \times G'(\mathbb{Q}_p)$ appearing in $L^2(\Gamma(N)\backslash G)$. Then ρ_p (which is an irreducible unitary representation of $G'(\mathbb{Q}_p) = PGL_2(\mathbb{Q}_p)$) is not from the complementary series of representations of $PGL_2(\mathbb{Q}_p)$ unless it is one-dimensional.*

Proof. From the Strong Approximation Theorem it follows (see 6.3) that ρ can be extended to a representation ρ' of $G'(\mathbb{A})$ appearing in $L^2(G'(\mathbb{Q})\backslash G'(\mathbb{A}))$ such that $\rho' = \bigotimes_\ell \rho'_\ell$ where $\rho'_p = \rho_p$ and $\rho'_\infty = \rho_\infty$. Now, by the Jacquet-Langlands correspondence (6.2.1) there exists a subrepresentation $\tilde{\rho} = \bigotimes_\ell \tilde{\rho}_\ell$ of $L^2(PGL_2(\mathbb{Q})\backslash PGL_2(\mathbb{A}))$ such that $\tilde{\rho}_p = \rho'_p = \rho_p$ and $\tilde{\rho}_\infty$ is a discrete series representation of $PGL_2(\mathbb{R})$. It follows now from Deligne's Theorem (6.1.2) that $\tilde{\rho}_p$ (and hence ρ_p) is not a complementary series representation (unless $\tilde{\rho}_p$, and hence also $\tilde{\rho}$, is one-dimensional). □

7.2 The Banach-Ruziewicz problem for $n = 2, 3$

We have shown in (3.4.1) that an affirmative solution to the Banach-Ruziewicz problem for S^n would follow once we present a finitely generated subgroup Γ of $SO(n+1)$ whose representation on $L_0^2(S^n) = \{f \in L^2(S^n) \mid \int f d\lambda = 0\}$ does not weakly contain the trivial representation. We now show that the group $\Gamma = \Gamma(1)$ of Theorem 7.1.1 projected into $G'(\mathbb{R}) \simeq SO(3)$ satisfies this condition:

Theorem 7.2.1. *For $n = 2, 3$, Lebesgue measure is the only finitely additive, rotationally invariant measure of total measure one defined on the Lebesgue measurable subsets of S^n.*

Proof. In (2.2.12) it was proved that any such measure must be absolutely continuous with respect to the Lebesgue measure and hence defines an invariant mean on $L^\infty(S^n)$. To prove that this invariant mean is the Lebesgue integral, it suffices by (3.4.1) to present a group Γ acting on S^n in such a way that its action on $L_0^2(S^n)$ does not weakly contain the trivial representation ρ_0. Let $\Gamma = \Gamma(1)$ from (7.1.1), $\pi : \Gamma \to SO(3)$ its projection to $SO(3)$, $\tilde{\pi}$ its representation on $L^2(SO(3))$ and

$\tilde{\pi}_0$ is the restriction of $\tilde{\pi}$ to the invariant subspace $L_0^2(SO(3))$. If the action of Γ on $L_0^2(S^2)$ weakly contains the trivial representation, then so does $\tilde{\pi}_0$.

Denote $G_p = G'(\mathbb{Q}_p)$, $G_\infty = G'(\mathbb{R}) = SO(3)$, $\Gamma = G'(\mathbb{Z}[\frac{1}{p}])$ as before, embedded diagonally in $G_\infty \times G_p$, $\bar{\Gamma}$ is the projection of Γ to G_p. $\bar{\Gamma}$ is a cocompact lattice in G_p since G_∞ is compact and Γ is a cocompact lattice in $G_\infty \times G_p$. Denote $V = L^2(SO(3))$, and $(V, \tilde{\pi})$ is the representation of Γ on V given by $\tilde{\pi}(\gamma)(f)(x) = f(\gamma^{-1}x)$ for $x \in SO(3)$, $f \in V$ and $\gamma \in \Gamma \simeq \bar{\Gamma}$. Let (H, σ) be $\mathrm{Ind}_{\bar{\Gamma}}^{G_p}(V, \tilde{\pi})$.

Lemma 7.2.2. *$(H, \sigma) \simeq L^2(\Gamma \backslash G_\infty \times G_p)$ as G_p-representations,*

Proof. By the definition of induced representation, H is the space of functions $\varphi : G_p \to V$ satisfying $\varphi(\gamma g) = \tilde{\pi}(\gamma)\varphi(g)$ for $\gamma \in \bar{\Gamma}$, $g \in G_p$ and $\int_{\bar{\Gamma}\backslash G_p} \| \varphi(g) \|^2 < \infty$. So, if $x \in SO(3)$, φ satisfies $(\varphi(\gamma g))(x) = \varphi(g)(\gamma^{-1}x)$. With such a $\varphi \in H$, we associate the function $\tilde{\varphi} : G_\infty \times G_p \to \mathbb{C}$ given by $\tilde{\varphi}(g_\infty, g_p) = (\varphi(g_p))(g_\infty)$. It is easy to check that $\tilde{\varphi} \in L^2(\Gamma \backslash G_\infty \times G_p)$.

Conversely, with $\psi \in L^2(\Gamma \backslash G_\infty \times G_p)$, we associate the function $\bar{\psi} : G_p \to V$ defined by: $(\bar{\psi}(g_p))(g_\infty) = \psi(g_\infty, g_p)$.

It is not difficult to check that these maps define the desirable isomorphism. It is a G_p-isomorphism when we keep in mind that G_p acts on H by σ where: $(\sigma(g'_p)\varphi)(g_p)(g_\infty) = \varphi(g_p g'_p)(g_\infty)$. □

Now, recall from 3.1 that induction is a continuous map between the corresponding unitary duals. Thus, if $\tilde{\pi}_0$ weakly contains the trivial Γ-representation, then $L_0^2(\Gamma \setminus (G'(\mathbb{R}) \times G'(\mathbb{Q}_p)))$ weakly contains the trivial G_p-representation. If ρ_p is a G_p-subrepresentation of $L_0^2(\Gamma \setminus G_\infty \times G_p)$, then let $H(\rho_p)$ be the complete direct sum of all the G_p-subrepresentations isomorphic to ρ_p. This space is G_∞ invariant since G_∞ commutes with G_p. This means that if ρ_p appears as a G_p-subrepresentation then there exists some representation of G_∞, say ρ_∞ such that $\rho_\infty \otimes \rho_p$ is a $G_\infty \times G_p$ subrepresentation of $L^2(\Gamma \backslash G_\infty \times G_p)$. By (7.1.1) this implies that ρ_p is not in the complementary series unless ρ_p is one-dimensional. Now if ρ_p is a one-dimensional representation, then it can not be the trivial representation since $\rho_p|_{\bar{\Gamma}}$ does not (strongly) contain the trivial representation; so it must be the sg representation (see remark after Corollary (5.5.3)). This representation is also bounded away from the trivial representation, and the theorem is proved for $n = 2$.

The above proof implies the theorem also for S^3. One way to argue is that $SO(4)$ is locally isomorphic to $SO(3) \times SO(3)$ and we have found a dense group Γ of $SO(3)$ whose representation in $L_0^2(SO(3))$ does not weakly contain the trivial representation. The same is therefore true for $\Gamma \times \Gamma$ in $SO(3) \times SO(3)$ and hence in $SO(4)$. Another approach is to prove Theorem 7.1.1 with $SU(2)$ instead of $SO(3)$ – see Drinfeld [Dr1] – and to use the fact that $SU(2)$, being homeomorphic to S^4, acts on S^4 transitively, so the same Γ which works for $SU(2)$ works for

S^3 and S^4. Anyway, the theorem is now proved. Moreover, it proves also the assertion promised in 3.4: Even though $K = SO(3)$ and $K = SO(4)$ do not contain any infinite group with property (T), they do contain a subgroup Γ whose representation on $L_0^2(K)$ does not weakly contain the trivial representation. The Haar measure is, therefore, the only finitely additive invariant measure of K for those K's as well! (But not for $SO(2) \simeq S^1$!) □

Before ending this section we mention that this affirmative solution for the Ruziewicz problem for $n = 2$ and 3 can also be deduced from Selberg's Theorem (4.4.1) (see also (5.5.2)) instead of (6.1.2). This will be done by choosing D to be a quaternion algebra over a real quadratic extension k of $\mathbb{Q}$. Pick D which splits at one infinite place and ramifies at the other. If $\mathbb{O}$ is the ring of integers of k, then $\Gamma = G'(\mathbb{O})$ is a lattice in $G'(k_{\infty_1}) \times G'(k_{\infty_2}) = PGL_2(\mathbb{R}) \times SO(3)$ where $G' = D^*/Z(D^*)$. Now, the Jacquet-Langlands Theorem (6.2.1) with Selberg's Theorem implies that the representation of Γ in $L_0^2(SO(3))$ does not weakly contain the trivial representation. This, as before, concludes the proof.

7.3 Ramanujan graphs and their extremal properties

Continuing with the notations set up at the beginning of the chapter, recall that the combinatorial Laplacian Δ of a k-regular graph X is $\Delta = kI - \delta$ where δ is the adjacency matrix, i.e., $\delta : L^2(X) \to L^2(X)$ is defined as $(\delta f)(x) = \sum_{\{y|d(y,x)=1\}} f(y))$.

Theorem 7.3.1. *Let $\Gamma(N)$ be as before and $\bar{\Gamma}(N)$ its projection into $G'(\mathbb{Q}_p) \simeq PGL_2(\mathbb{Q}_p)$ under the map $G'(\mathbb{R}) \times G'(\mathbb{Q}_p) \to G'(\mathbb{Q}_p)$. Then*

$$X = \bar{\Gamma}(N)\backslash PGL_2(\mathbb{Q}_p)/PGL_2(\mathbb{Z}_p)$$

is a finite $(p+1)$-regular graph satisfying: every eigenvalue λ of $\delta = \delta_X$ is either $\pm(p+1)$ or $|\lambda| \le 2\sqrt{p}$. So, X is a Ramanujan graph.

Proof. First, note that X is indeed a graph since it is a quotient of the tree $PGL_2(\mathbb{Q}_p)/PGL_2(\mathbb{Z}_p)$ by the discrete group $\bar{\Gamma}(N)$. It is finite since $\bar{\Gamma}(N)$ is cocompact in $PGL_2(\mathbb{Q}_p)$ and so X is discrete and compact and hence finite. It is $(p+1)$-regular since the tree is so (X might have multiple edges for small values of N).

To prove the assertion on λ, we use (5.5.3). By that corollary, it suffices to show that no complementary series representation occurs in $L^2(\bar{\Gamma}(N)\backslash PGL_2(\mathbb{Q}_p))$. If such a representation ρ occurs there, then $\rho_0 \otimes \rho$ occurs as a subrepresentation of $L^2(\Gamma(N)\backslash SO(3) \times PGL_2(\mathbb{Q}_p))$ where ρ_0 is the trivial representation of $SO(3)$. This contradicts Theorem 7.1.1 and our theorem is therefore proven. □

The rest of this section will be devoted to proving that the family of graphs X defined in Theorem (7.3.1) satisfy some more extremal properties. We begin with the girth. To this end we start with some lemmas.

Lemma 7.3.2. *Let $g \in SL_2(\mathbb{Q}_p)$ and x a vertex in the tree $T = PGL_2(\mathbb{Q}_p)/PGL_2(\mathbb{Z}_p)$. Then $d(gx, x) \geq -2v_p(Tr(g))$, where ν_p is the p-adic valuation and $d(x, y)$ denotes the distance between two vertices x and y.*

Proof. The action of $GL_2(\mathbb{Q}_p)$ on the tree T preserves the distance between vertices and $d(gx, x) = d(hgx, hx) = d(hgh^{-1}(hx), hx)$ for any $h \in GL_2(\mathbb{Q}_p)$. On the other hand, $GL_2(\mathbb{Q}_p)$ acts on the vertices of T transitively, and the trace of a matrix is preserved under conjugation. Therefore, substituting x by x_0 – the standard lattice – (see 5.3) and g by hgh^{-1} where $hx = x_0$, we may assume $x = x_0$. So we have to show that for any $g \in SL_2$, $d(gx_0, x_0) \geq -2v_p(Tr(g))$.

For $g = \begin{pmatrix} a_1 & a_2 \\ a_3 & a_4 \end{pmatrix} \in GL_2(\mathbb{Q}_p)$ denote $w(g) = \max_{i=1,\dots,4}\{-v_p(a_i)\}$. From the ultra-metric inequality, it follows that

$$w(g) \geq -v_p(a_1 + a_4) = -v_p(Tr(g)).$$

So, it suffices to prove: $d(gx_0, x_0) \geq 2w(g)$. We will, in fact, prove equality:

$$d(gx_0, x_0) = 2w(g). \qquad (*)$$

Let Y be the subset of all $g \in SL_2(\mathbb{Q}_p)$ for which $(*)$ holds. Since v_p is a non-archimedean valuation, $w(gh) \leq w(g) + w(h)$ for any $g, h \in GL_2(\mathbb{Q}_p)$. But $w(h) = 0$ for any $h \in GL_2(\mathbb{Z}_p)$. Therefore, $w(pgp') = w(g)$ for any $g \in GL_2(\mathbb{Q}_p)$ and $p, p' \in GL_2(\mathbb{Z}_p)$. On the other hand, since $GL_2(\mathbb{Z}_p)(x_0) = x_0$, $d(pgp'x_0, x_0) = d(gx_0, x_0)$ for any $g \in GL_2(\mathbb{Q}_p)$ and $p, p' \in GL_2(\mathbb{Z}_p)$. Therefore

$$GL_2(\mathbb{Z}_p)Y\, GL_2(\mathbb{Z}_p) = Y.$$

Let $\gamma = \begin{pmatrix} p^{-1} & 0 \\ 0 & p \end{pmatrix}$. Then a direct computation shows that for any $n \geq 0$, $\operatorname{dist}(\gamma^n x_0, x_0) = 2n$ and $w(\gamma^n) = n$, so $A^+ = \{\gamma^n \mid n \geq 0\}$ is also in Y. By the Cartan decomposition, $SL_2(\mathbb{Q}_p) = SL_2(\mathbb{Z}_p) \cdot A^+ \cdot SL_2(\mathbb{Z}_p)$ (prove directly!). Hence, $Y = SL_2(\mathbb{Q}_p)$ and the lemma is proven. $\square$

Remark 7.3.3. (i) The lemma is not valid for all $g \in GL_2(\mathbb{Q}_p)$ simply because the scalars act trivially on T.

(ii) It is possible to show that the lemma cannot be improved in the following sense: For any $g \in SL_2(\mathbb{Q}_p)$ there exists a vertex $x \in T$ such that $\operatorname{dist}(gx, x) = \max\{0, -2v_p(Tr(g))\}$. The details are left to the reader.

Definition 7.3.4. *If X is a graph, the girth $g(X)$ of X is the length of the shortest non-trivial closed path in X.*

Lemma 7.3.5. *Let T be a tree on which a group Γ acts freely (i.e., $\forall 1 \neq \gamma \in \Gamma$ and $x \in T$, $\gamma x \neq x$) with quotient graph X. Then* $g(X) = \min_{\substack{1 \neq \gamma \in \Gamma \\ x \in T}} \operatorname{dist}(\gamma x, x)$.

Proof. Let $x' \in T$ and $1 \neq \gamma' \in \Gamma$ be such that the minimum obtained is $\operatorname{dist}(\gamma' x', x')$. Then the path in T from x' to $\gamma' x'$ gives a non-trivial closed path in X. Conversely, every closed path in X can be lifted up to a (non-closed) path in T such that its endpoints are in the same orbit; this shows that $g(X) \geq \min \operatorname{dist}(\gamma x, x)$. □

Before continuing we clear up a technical point by showing that for «many» N-s, $\Gamma(N)$ satisfies both assumptions of the last two lemmas, i.e., it is in $PSL_2(\mathbb{Q}_p)$ and it is torsion free, and so its action on the tree $T = PGL_2(\mathbb{Q}_p)/PGL_2(\mathbb{Z}_p)$ is a free action.

Lemma 7.3.6. *Let $D = D(u, v)$ where $u, v > 0$ and assume that p is not a quadratic residue* $\operatorname{mod} N$ *and let $G' = D^*/Z(D^*)$ as before. Then* $\Gamma(N) = \ker(G'(\mathbb{Z}[\frac{1}{p}]) \to G'(\mathbb{Z}[\frac{1}{p}]/N\mathbb{Z}[\frac{1}{p}]))$ *is in $PSL_2(\mathbb{Q}_p)$, it is torsion free and hence a free group which acts freely on T.*

Proof. If $\gamma \in \Gamma(1)$ then $\det(\gamma)$ (when we consider γ as an element of $D(\mathbb{Z}[\frac{1}{p}])$ before factoring by the center) is an invertible element of $\mathbb{Z}[\frac{1}{p}]$ and $\det(\gamma) = \gamma\bar{\gamma} > 0$ (since $u, v > 0$). $\det(\gamma)$ is therefore in $U = \{p^n \mid n \in \mathbb{Z}\}$. If $\gamma \in \Gamma(N)$ then $\det(\gamma) \equiv 1 \pmod{N}$. We claim that $\det(\gamma) \in U^2 = \{p^{2n} \mid n \in \mathbb{Z}\}$. Otherwise $\det(\gamma) = p^{2r+1}$ and so $(p^{r+1})^2 \equiv p \pmod{N}$, which contradicts our assumption on p. So $\det(\gamma)$ is a square in $\mathbb{Z}[\frac{1}{p}]$ and hence in $\mathbb{Q}_p$, and so $\gamma \in PSL_2(\mathbb{Q}_p)$. $\Gamma(N)$ is also torsion free: Let q be a prime dividing N; then $\Gamma(1)$ is embedded in $G'(\mathbb{Z}_q)$ in such a way that $\Gamma(N) \hookrightarrow L_q = \ker(G'(\mathbb{Z}_q) \to G'(\mathbb{Z}_q/q\mathbb{Z}_q))$. Now, L_q is a torsion-free group. This can be deduced from the logarithmic map convergence on L_q. (See also [DDMS, Chapter 5]. An easy proof of the torsion freeness of $\Gamma(N)$ in case two primes q_1 and q_2 divide N, is: L_{q_i} is a pro-q_i group so the only possible torsion in $\Gamma(N)$ is q_i-torsion. Since $q_1 \neq q_2$ this means that there is no torsion.)

Once we know that $\Gamma(N)$ is torsion free, plus the fact that it is discrete in $PGL_2(\mathbb{Q}_p)$, it follows that it acts freely on the tree T and hence is a free group (cf. [S1, p. 82]). □

We are now ready to prove the following:

Theorem 7.3.7. *Using all notation and all assumptions above (so $u, v > 0$ and p is <u>not</u> quadratic residue* $\operatorname{mod} N$*), the girth $g(X)$ of $X = \Gamma(N)\backslash PGL_2(\mathbb{Q}_p)/PGL_2(\mathbb{Z}_p)$ is at least* $4\log_p(N) - 4\log_p(2)$.

Proof. By Lemma 7.3.5, $g(X)$ can be estimated by $\mathrm{dist}(\gamma x, x)$ for every $\gamma \in \Gamma(N)$ and $x \in T$. Lemma 7.3.2 (with the help of (7.3.6)) yields that we can estimate $-2v_p(Tr(\gamma))$ instead (where γ is replaced by a representative of it, called γ, in $(\mathbb{Z}[\frac{1}{p}] \underset{\mathbb{Z}}{\otimes} D)^*$ whose norm satisfies $\gamma\bar{\gamma} = 1$).

Well, $\gamma = a + bi + cj + dk$ where $a, b, c, d \in \mathbb{Z}[\frac{1}{p}]$, $1 = \gamma\bar{\gamma} = a^2 + ub^2 + vc^2 + uvd^2$, $Tr(\gamma) = \gamma + \bar{\gamma} = 2a$ and $\gamma \equiv 1 \pmod N$, i.e., $a \equiv 1 \pmod N$ and $b \equiv c \equiv d \equiv 0 \pmod N$. Now, $2 - Tr(\gamma) = 2 - 2a = a^2 + ub^2 + vc^2 + uvd^2 + 1 - 2a = (a-1)^2 + ub^2 + vc^2 + uvd^2$. Thus $2 - Tr(\gamma)$ is divisible (in $\mathbb{Z}[\frac{1}{p}]$) by N^2. Write $2 - Tr(\gamma) = \frac{N^2 m}{p^\ell}$ where $(m, p) = 1$ and $m, \ell \in \mathbb{Z}$. Since γ is not trivial in $(\mathbb{Z}[\frac{1}{p}] \underset{\mathbb{Z}}{\otimes} D)^* / Z((\mathbf{Z}[\frac{1}{p}] \underset{\mathbb{Z}}{\otimes} D)^*)$, at least one of the scalars b, c or d is not zero. It is impossible, therefore, that $a = 1$ (which would imply that $ub^2 + vc^2 + uvd^2 = 0$, i.e., the quadratic form associated with D represents 0 non-trivially over $\mathbb{Q}$; this is impossible by our assumption). So $Tr(\gamma) \neq 2$ and hence $m \neq 0$. But, since $a^2 + ub^2 + vc^2 + uvd^2 = 1$ and we assume $u, v > 0$, we have $|a| \leq 1$. Therefore,

$$|\frac{N^2 m}{p^\ell}| = |2 - 2a| \leq 4.$$

From this, we get $p^\ell \geq |\frac{N^2 m}{4}| \geq \frac{N^2}{4}$ and hence $\ell \geq \log_p(\frac{N^2}{4}) = 2\log_p(N) - 2\log_p(2)$. Moreover, $Tr(\gamma) = 2 - \frac{N^2 m}{p^\ell} = \frac{2p^\ell - N^2 m}{p^\ell}$ and $(p, 2p^\ell - N^2 m) = 1$ so $v_p(Tr(\gamma)) = -\ell$. Lemma 7.3.2 now concludes the proof of the theorem. □

To appreciate the importance of the above theorem we add a few remarks:

Remarks 7.3.8. (i) The number of vertices of the graph X in the theorem can be estimated in the following way: Let $h = |\Gamma(1) \backslash PGL_2(\mathbb{Q}_p)/PGL_2(\mathbb{Z}_p)|$. Since $\Gamma = \Gamma(1)$ is cocompact, h is finite. (In fact, h is a very interesting number from an arithmetic point of view. It is the «class number» of the quaternion algebra – see [Kn] and [Vi 1, p. 87].) Since $\Gamma(1)/\Gamma(N)$ is a subgroup of $G'(\mathbb{Z}/N\mathbb{Z})$, $[\Gamma(1) : \Gamma(N)] \leq N^3$ so $|X| \leq hN^3$. Hence, if we fix D (and hence h; moreover we can easily find D for which $h = 1$; see 7.4), then as N varies one obtains an infinite family of $(p+1)$-regular graphs X_i with girth $(X_i) \geq \frac{4}{3}\log_p(|X_i|) - 4\log_p(2)$. (By a more careful analysis one can omit the $4\log_p(2)$ term, which is negligible anyway.)

(ii) A simple counting argument shows that for a k-regular graph X, $g(X) \leq (2 + o(1))\log_{k-1}(|X|)$ when $|X| \to \infty$. Erdös and Sachs [ES] proved using random considerations that graphs with girth $\log_{k-1}(|X|)$ exist. Thus the above (explicit!) graphs are better than those obtained by random methods! (For the case $k = 3$, Weiss [We] has previously proved that the graphs constructed by Biggs and Hoare in [BH] also satisfy $g(x) \geq \frac{4}{3}\log_2(|X|)$). It is still an open problem whether for a fixed k, an infinite family of k-regular graphs X_i exist with $\limsup \frac{g(X_i)}{\log_{k-1}(|X_i|)} > \frac{4}{3}$. (See [B1,2,3] for more on this problem.)

(iii) Instead of making the assumption that p is not a quadratic residue mod N, we could replace $\Gamma(N)$ by $\Gamma_1(N) = \Gamma(N) \cap PSL_2(\mathbb{Q}_p)$ and the proof would work word for word for the graphs $X = \Gamma_1(N)\backslash T$. Note that for groups Γ in $PSL_2(\mathbb{Q}_p)$, Γ preserves the bi-partition of the tree. (Every tree is bi-partite by declaring two vertices to be «on the same side» if the distance between them is even. By the proof of Lemma (7.3.2), every element of $PSL_2(\mathbb{Q}_p)$ preserves this partition.) Hence $\Gamma\backslash T$ is bi-partite. Bi-partite graphs can be colored by two colors (see Definition (7.3.10) below). Our next goal will be to get graphs with high chromatic number – so then it will be important that $\Gamma(N)$ is <u>not</u> in $PSL_2(\mathbb{Q}_p)$. We can still assert that girth $(\Gamma_1(N)\backslash T) \geq 4\log_p(N) - 4\log_p(2)$. If $\Gamma_1(N) \neq \Gamma(N)$ then $[\Gamma(N) : \Gamma_1(N)] = 2$, and so $\Gamma_1(N)\backslash T$ is a double cover of $\Gamma(N)\backslash T$. From all this, it follows that if p is a quadratic residue mod N we still have the weaker result: girth $(\Gamma(N)\backslash PGL_2(\mathbb{Q}_p)/PGL_2(\mathbb{Z}_p)) \geq 2\log_p(N) - 2\log_p(2)$.

Lemma 7.3.9. *Using the notation above, now assume that $p \neq 2$ and p <u>is</u> a quadratic residue* mod N. *Then* $X = \Gamma(N)\backslash PGL_2(\mathbb{Q}_p)/PGL_2(\mathbb{Z}_p)$ *is* <u>not</u> *bi-partite.*

Proof. X is bi-partite if and only if $-(p+1)$ is an eigenvalue of δ the adjacency matrix of X. By (5.5.1) and the last paragraph of 5.4, this happens if and only if the representation sg of $PGL_2(\mathbb{Q}_p)$ is a subrepresentation of $L^2(\Gamma(N)\backslash PGL_2(\mathbb{Q}_p))$. The representation sg is the non-trivial one-dimensional representation of $PGL_2(\mathbb{Q}_p)$, which is trivial on the index two subgroup $PSL_2(\mathbb{Q}_p) \cdot PGL_2(\mathbb{Z}_p)$ (recall that $|PGL_2(\mathbb{Q}_p)/PSL_2(\mathbb{Q}_p)| = |\mathbb{Q}_p^*/(\mathbb{Q}_p^*)^2| = 4$). As p is a quadratic residue mod N, $\Gamma(N)$ contains elements γ with $\det(\gamma) = p^{2r+1}$ for some r. (See the proof of Lemma 7.3.6.) Hence $\Gamma(N) \underset{\neq}{\subset} PSL_2(\mathbb{Q}_p) \cdot PGL_2(\mathbb{Z}_p)$ and hence $\Gamma(N) \cdot PSL_2(\mathbb{Q}_p) \cdot PGL_2(\mathbb{Z}_p) = PGL_2(\mathbb{Q}_p)$. This implies that sg cannot appear in $L^2(\Gamma(N)\backslash T)$, and X is, therefore, not bi-partite. □

The results on the graphs discussed above can be applied to estimate some more combinatorial invariants which are related to eigenvalues. As a sample we will mention some. (For more combinatorial invariants related to eigenvalues, see [Bg] and [CDS].)

Definition 7.3.10. *Let X be a graph.*

(i) $\chi(X)$ – *the chromatic number of X is the smallest number of colors needed to color the vertices of X in such a way that adjacent vertices have different colors.*

(ii) Diam(X) – *the diameter of X is* max dist(x, y) *when x and y run over the vertices of X.*

(iii) $i(X)$ – *the independence number of X is the maximal size of a subset of X with no two vertices in the subset adjacent to each other.*

All these invariants can be related to eigenvalues in the following way:

Let's set up some notation: for a connected k-regular graph on n vertices $X = X_{n,k}$, the eigenvalues of δ, the adjacency matrix of X, are $\lambda_0 = k > \lambda_1 \geq \ldots \geq \lambda_{n-1}$. X is bi-partite iff $\lambda_{n-1} = -k$. Let $\lambda_1(X) = \lambda_1$ and $\lambda(X)$ be defined as $\max\{|\lambda_1|, |\lambda_{n-2}|\}$ if X is bi-partite and $\max\{|\lambda_1|, |\lambda_{n-1}|\}$ if not. So X is a Ramanujan graph if and only if $\lambda(X) \leq 2\sqrt{k-1}$. Finally, ℓn is the natural logarithm.

Proposition 7.3.11. *I. Let $X = X_{n,k}$ be a k-regular graph on n vertices. Then*

(i) $\chi(X) \geq \frac{k}{-\lambda_{n-1}(X)} + 1$.

(ii) $\mathrm{Diam}(X) \leq \frac{\ell n(2|X|)}{\ell n(\frac{k+\sqrt{k^2-\lambda(X)^2}}{\lambda(X)})}) \leq \frac{\ell n(2|X|)}{\ell n(\frac{k}{\lambda(X)})}$.

(iii) $i(X) \leq \frac{\lambda(X)}{k}|X|$.

II. If X is a k-regular Ramanujan graph, then:

(i) $\chi(X) = 2$ *if X is bi-partite and* $\chi(X) \geq \frac{k}{2\sqrt{k-1}} + 1$ *if not.*

(ii) $\mathrm{Diam}(X) \leq 2\log_{k-1}(|X|) + \log_{k-1}(4)$.

(iii) $i(X) \leq \frac{2\sqrt{k-1}}{k}|X|$.

Proof. I. (i) is due to Hoffman [Ho] (or follows from (iii) without the «1», since $\chi(X) \geq \frac{|X|}{i(X)}$).

(ii) is proved in [LPS2, Theorem 5.1] or [Sa2, 3.2.6] (in somewhat weaker forms it is also proved in [AM] and [C2]).

(iii) A proof provided by N. Alon is given in [LPS2, Proposition 5.2], and one due to Hoffman appears in Haemers [H].

Part II follows from Part I by simple substitutions. □

Applying the last proposition to the graphs studied above provide the following theorem, which summarizes the results of this section:

Theorem 7.3.12. *Let $D = D(u, v)$ be a definite quaternion algebra, ($0 < u, v \in \mathbb{Q}$ and $i^2 = -u$, $j^2 = -v$ and D ramifies at ∞). Let G' be the $\mathbb{Q}$-algebraic group $G = D^*/Z(D^*)$, p a prime for which D splits over $\mathbb{Q}_p$, N an integer prime to p, $\Gamma(N) = \ker(G'(\mathbb{Z}[\frac{1}{p}]) \to G'(\mathbb{Z}[\frac{1}{p}]/N\mathbb{Z}[\frac{1}{p}]))$ and $T = PGL_2(\mathbb{Q}_p)/PGL_2(\mathbb{Z}_p)$ the $(p+1)$-regular tree. Let $X = X^{p,N}$ be the $(p+1)$-regular finite graph $X = \Gamma(N)\backslash T$. Then X is a Ramanujan graph with less than N^3 vertices.*

Moreover: I. If p is not a quadratic residue $\bmod N$, *then:*

(i) *X is bi-partite, $\chi(X) = 2$ and $i(X) = \frac{|X|}{2}$.*

(ii) *girth $(X) \geq 4\log_p(N) - 4\log_p(2) \simeq \frac{4}{3}\log_p(X)$.*

(iii) *$Diam(X) \leq 2\log_p(|X|) + \log_p(4)$.*

II. If p is a quadratic residue mod N, *then*

(i) X *is not bi-partite.*

(ii) $i(X) \leq \frac{2\sqrt{p}}{p+1}|X|$.

(iii) $\chi(X) \geq \frac{p+1}{2\sqrt{p}} + 1$.

(iv) *girth* $(X) \geq 2\log_p N - 2\log_p(2) \simeq \frac{2}{3}\log_p(N)$.

(v) *Diam*$(X) \leq 2\log_p(|X|) + \log_p(4)$.

Some remarks are in order: The question of existence of graphs with large girth and large chromatic number was a long outstanding open problem. The difficulty comes from the fact that if X has large girth it «locally» looks like a tree, and a tree can always be colored by two colors. Erdös was the first one to prove the existence of such graphs using random methods (see [B1,2] for references and the history of the problem). Our explicit graphs have the stronger property of large girth and $\frac{i(X)}{|X|}$ arbitrarily small (this implies $\chi(X)$ is arbitrarily large since clearly $\chi(X) \geq \frac{|X|}{i(X)}$). It was shown recently by Biggs [Bg2] that the lower bounds given in the theorem for the girth are tight.

Regarding the diameter: It is a simple exercise to prove that for a k-regular graph Diam$(X) \geq \log_{k-1}(|X|)-2$, but it is not known whether asymptotic equality is possible for infinitely many graphs. See [B1, B2] for more information on this problem.

The question of existence of k-regular Ramanujan graphs for most k is still open. In [Mor1], M. Morgenstern considers the case where k is of the form $k = p^\alpha + 1$, where p a prime. For this case he uses global fields in characteristic $p > 0$ to construct graphs analogous to those here. He uses the work of Drinfeld [Dr2], which is the characteristic p analogue of Ramanujan conjecture (see more in Theorem 7.4.5 and in §8.4). But for $k \neq p^\alpha + 1$ nothing seems to be known. The first open case is $k = 7$.

7.4 Explicit constructions

The graphs studied above, $X = X^{p,N}$, were explicit in the sense that we presented specific graphs and did not merely prove their existence. Still, their construction was not explicit since they were presented as quotients of an infinite tree by an infinite group. For various reasons (the main one being the applicability of these graphs as excellent expanders (by 4.2) which can serve as components of various communication networks (see 1.1)), it is desirable to have explicit constructions. This will be our main goal in this section. We follow the construction in Lubotzky-Phillips-Sarnak [LPS 1, LPS2] but from a different point of view.

We give the explicit construction for the case $D = D(1, 1)$ and $p \equiv 1 \pmod 4$. Under these assumptions $D_{\mathbb{Q}_p}$ splits, and the following is an explicit isomorphism

$\sigma : D_{\mathbb{Q}_p} \to M_2(\mathbb{Q}_p)$,

$$\sigma(x_0 + x_1 i + x_2 j + x_3 k) = \begin{pmatrix} x_0 - x_1\varepsilon & x_2 + x_3\varepsilon \\ -x_2 + x_3\varepsilon & x_0 - x_1\varepsilon \end{pmatrix}$$

where $\varepsilon \in \mathbb{Q}_p$ is an element satisfying $\varepsilon^2 = -1$.

Let $S' = \{(x_0, x_1, x_2, x_3) \in \mathbb{Z} \mid \sum_{i=1}^{4} x_i^2 = p\}$. By Jacobi's theorem (2.1.8), $|S'| = 8(p+1)$. Since $p \equiv 1 \pmod 4$, for every $(x_i) \in S'$ one of the x_i's is odd and the others are even. Let S be the subset $S = \{(x_0, x_1, x_2, x_3) \in S' \mid x_0 > 0$ and odd and x_i is even for $i = 1, 2, 3\}$. Then $|S| = p + 1$. To each element $(x_0, x_1, x_2, x_3) \in S$ we associate the quaternion $\alpha = x_0 + x_1 i + x_2 j + x_3 k$, so $\alpha\bar{\alpha} = p$. We denote also by S this set of $p + 1$ quaternions. S can be written as $\{\alpha_1, \dots \alpha_s, \bar{\alpha}_1, \dots, \bar{\alpha}_s\}$ where $s = \frac{p+1}{2}$ since $\alpha \in S$ iff $\bar{\alpha} \in S$. Each one of the elements in S is invertible in $D_{\mathbb{Z}[\frac{1}{p}]}$ (since $\alpha\bar{\alpha} = p$), and so they give, after projection modulo the center, elements in $\Gamma = G'(\mathbb{Z}[\frac{1}{p}])$. The assumptions made on the x_i's actually imply that $S \in \Gamma(2)$ – the congruence subgroup modulo 2. Now recall the group $\Lambda(2)$ we considered in (2.1.11): a close look shows that it is exactly our $\Gamma(2)$. (There we were interested in its embedding as a dense subgroup in $G'(\mathbb{R}) \simeq SO(3)$; now we are interested in its embedding as a discrete subgroup in $G'(\mathbb{Q}_p) \simeq PGL_2(\mathbb{Q}_p)$.) Therefore by (2.1.11), it is a free group on the $s = \frac{p+1}{2}$ generators $\alpha_1, \dots, \alpha_s$ and $\bar{\alpha}_i = \alpha_i^{-1}$. Considering $\Gamma(2)$ as a subgroup of $PGL_2(\mathbb{Q}_p)$ (under the isomorphism σ defined above), it is a cocompact lattice. In fact, we have the following

Lemma 7.4.1. *The action of $\Gamma(2)$ on the tree $T = G/K = PGL_2(\mathbb{Q}_p)/PGL_2(\mathbb{Z}_p)$ is simply transitive. Hence $G = \Gamma(2) \cdot K$, i.e., every $g \in G$ can be written in a* unique *way as $g = \gamma \cdot k$ for some $\gamma \in \Gamma(2)$ and $k \in K$.*

Proof. Look at the standard lattice $L_0 \simeq \mathbb{Z}_p \times \mathbb{Z}_p$ in $V = \mathbb{Q}_p \times \mathbb{Q}_p$ (see 5.3). Since $\det(\alpha_i) = p$ for every $i = 1, \dots, p+1$, α_i takes L_0 to a sublattice of L_0 of index p, i.e., to a neighbor of $x_0 = [L_0]$ as an element of the tree T. The vertices $\alpha_1(x_0), \dots, \alpha_{p+1}(x_0)$ must be all different since if $\alpha_i(x_0) = \alpha_j(x_0)$ then $\alpha_i^{-1}\alpha_j(x_0) = x_0$, i.e., $\alpha_i^{-1}\alpha_j$ is in the compact stabilizer of x_0 and also in the discrete subgroup $\Gamma(2)$ and hence is an element of finite order; but $\Gamma(2)$ is free whence torsion free and so $\alpha_i^{-1}\alpha_j = 1$ and $\alpha_i = \alpha_j$. This proves that $\alpha_1(x_0), \dots, \alpha_{p+1}(x_0)$ are exactly all the $p+1$ neighbors of x_0. It is not difficult to deduce from this that the action of $\Gamma(2)$ on T is transitive. It is a free action (i.e., no $1 \neq \gamma \in \Gamma(2)$ has any fixed point) by the argument above. □

From our discussion in the proof of the lemma we see that T can be identified with the Cayley graph of $\Gamma(2)$ with respect to the set S of generators. This last observation is crucial for us: Let $N = 2M$ where M is odd and N is prime to p. Then the congruence subgroup $\Gamma(N)$ is a normal subgroup of $\Gamma = \Gamma(1)$

and in particular of $\Gamma(2)$. The graph we are interested in is $\Gamma(N)\backslash T$, but by the above identification it is indeed the Cayley graph of the quotient finite group $\Gamma(N)\backslash\Gamma(2) = \Gamma(2)/\Gamma(N)$ with respect to the set of generators S. The only thing we were left with is recognizing the finite group $\Gamma(2)/\Gamma(N)$. By its definition it is (isomorphic to) a subgroup of $G'(\mathbb{Z}/N\mathbb{Z}) \simeq G'(\mathbb{Z}/2\mathbb{Z}) \times G'(\mathbb{Z}/M\mathbb{Z})$. The image of $\Gamma(2)$ in the first direct factor is trivial, so we can identify $\Gamma(2)/\Gamma(N)$ with a subgroup of $G'(\mathbb{Z}/M\mathbb{Z})$. This last group is isomorphic to $PGL_2(\mathbb{Z}/M\mathbb{Z})$. If -1 is a quadratic residue mod M, it is also not difficult to give an explicit isomorphism.

Lemma 7.4.2. *The image of* $\Gamma(2)$ *in* $PGL_2(\mathbb{Z}/M\mathbb{Z})$ *contains* $PSL_2(\mathbb{Z}/M\mathbb{Z})$.

Proof. Let H be the $\mathbb{Q}$-algebraic group of the elements of D of norm 1. This group satisfies the assumptions of the Strong Approximation Theorem. (Indeed, $H(\mathbb{C}) \simeq SL_2(\mathbb{C})$ and so H is simply connected as an algebraic group; see also [Vi 1, Theorem 4.3, p. 81].) Hence (see 6.3) $H(\mathbb{Q})\cdot H(\mathbb{Q}_p)$ is dense in $H(\mathbb{A})$. Dividing both sides by $H(\mathbb{Q}_p)$ we deduce that $H(\mathbb{Q})$ is dense in $F = H(\mathbb{R}) \cdot \prod'_{p\neq\ell<\infty} H(\mathbb{Q}_\ell)$. Let K be the following compact open subgroup of F; $K = H(\mathbb{R}) \times \prod_{p\neq\ell<\infty} K_\ell$ when $K_2 = \ker(H(\mathbb{Z}_2) \to H(\mathbb{Z}_2/2\mathbb{Z}_2))$ and $K_\ell = H(\mathbb{Z}_\ell)$ for every $\ell \neq 2, p$. Then $H(\mathbb{Q}) \cap K$ is dense in K. A careful (but easy) check shows that $H(\mathbb{Q}) \cap K$ is equal to the congruence subgroup mod 2 of the arithmetic group $\Gamma_1 = H(\mathbb{Z}[\frac{1}{p}])$. So $\Gamma_1(2)$ is dense in K and hence its projection is onto any finite quotient of K. In particular since $K/H(\mathbb{R})$ is the congruence subgroup mod 2 of $H(\prod_{\ell\neq p} \mathbb{Z}_\ell)$, K and hence $\Gamma_1(2)$ is mapped epimorphically on $H(\mathbb{Z}/M\mathbb{Z})$. This last group is isomorphic to $SL_2(\mathbb{Z}/M\mathbb{Z})$ by our assumptions on M.

The group H is mapped into G' by dividing H by its center. The map $H \to G'$ is a subjective map as a map between algebraic groups (i.e., over $\mathbb{C}$) but usually not for subrings or subfields of $\mathbb{C}$. Γ_1 is mapped into Γ and the image is a subgroup of index 2. (Note that $\alpha_i \in \Gamma$ are not in the image of Γ_1 since $\det(\alpha_i) = p$ and p is not a square in $\mathbb{Z}[\frac{1}{p}]$.) Similarly, the image of $\Gamma_1(2)$ is an index two subgroup in $\Gamma(2)$. The projection $\Gamma_1(2) \to \Gamma(2) \to \Gamma(2)/\Gamma(N) \to PGL_2(\mathbb{Z}/M\mathbb{Z})$ can be checked to be the same as $\Gamma_1(2) \to PGL_2(\mathbb{Z}/M\mathbb{Z})$ obtained before (through K).

All this shows that the image of $\Gamma(2)$ in $PGL_2(\mathbb{Z}/M\mathbb{Z})$ is a subgroup containing $PSL_2(\mathbb{Z}/M\mathbb{Z})$ as a subgroup of index ≤ 2. The question whether it is equal to $PSL_2(\mathbb{Z}/M\mathbb{Z})$ depends on whether the images of the α_i's are in $PSL_2(\mathbb{Z}/M\mathbb{Z})$. This will be the case if and only if $\det(\alpha_i) = p$ is a quadratic residue mod M (or equivalently mod $N = 2M$). □

To summarize we have the following theorem:

Theorem 7.4.3. *Let* $D = D(1,1)$ *be the Hamiltonian quaternion algebra,* $p \equiv 1 \pmod 4$ *a prime and* $N = 2M$ *such that* $(M, 2p) = 1$, *and assume that there exists* $\varepsilon \in \mathbb{Z}$ *such that* $\varepsilon^2 \equiv -1 \pmod M$. *To each of the* $p+1$ *solutions* $\alpha = (x_0, x_1, x_2, x_3)$ *of* $x_0^2 + x_1^2 + x_2^2 + x_3^2 = p$ *where* $x_0 > 0$ *is odd* *and x_1, x_2 and x_3 are even, associate the following matrix of $PGL_2(\mathbb{Z}/M\mathbb{Z})$, $\alpha \mapsto \begin{pmatrix} x_0 + x_1\varepsilon & x_2 + x_3\varepsilon \\ -x_2 + x_3\varepsilon & x_0 - x_1\varepsilon \end{pmatrix}$* (mod M). *Let L be the subgroup generated by the set S of those $(p+1)$ α's (note: $S = S^{-1}$). Then the Cayley graph of L with respect to S is isomorphic to $\Gamma(N)\backslash PGL_2(\mathbb{Q}_p)/PGL_2(\mathbb{Z}_p)$ when $\Gamma(N)$, as before, is the N-congruence subgroup of $G'(\mathbb{Z}[\frac{1}{p}])$ and G' the algebraic group $D^*/Z(D^*)$.*

Moreover, if p is a quadratic residue mod N, *then $L = PSL_2(\mathbb{Z}/M\mathbb{Z})$ and the Cayley graph $X(L;S)$ is a non-bi-partite Ramanujan graph satisfying all the claims of Theorem (7.3.12), Part II.*

If p is not a quadratic residue mod N, *then L contains $P = PSL_2(\mathbb{Z}/M\mathbb{Z})$ as an index two subgroup, $S \cap P = \phi$ and so $X(L;S)$ is a bi-partite Ramanujan graph satisfying all the claims of Theorem (7.3.12), Part I.*

Remarks 7.4.4. (i) Note that also from the explicit construction we see the dichotomy between the bi-partite case and the non-bi-partite case. Again this depends on p being a quadratic residue mod N.

(ii) The group L can be described precisely also when p is not a quadratic residue mod N: If $M = q^r$ is a prime power then $L = PGL_2(\mathbb{Z}/q^r\mathbb{Z})$ (assuming p is not a quadratic residue mod q). In general write $M = \prod_{i=1}^{d_1} q_i^{r_i} \cdot \prod_{i=d_1+1}^{d_1+d_2} q_i^{r_i}$ where $\{q_1, \ldots, q_{d_1}\}$ are the prime factors of M for which p is not a quadratic residue and $\{q_{d_1+1}, \ldots, q_{d_1+d_2}\}$ are those for which p is quadratic residue. Then $PGL_2(\mathbb{Z}/M\mathbb{Z})/PSL_2(\mathbb{Z}/M\mathbb{Z}) \simeq \{\pm 1\}^{d_1} \times \{\pm 1\}^{d_2}$ in a natural way. L is the pre-image of the diagonal subgroup of $\{\pm 1\}^{d_1}$. One side of this bi-partite graph (in case $d_1 > 0$) is $PSL_2(\mathbb{Z}/M\mathbb{Z})$ the pre-image of $(1, \ldots, 1)$, and the other side is the pre-image of $(-1, \ldots, -1, 1, \ldots, 1)$.

(iii) For $N = 2q$, q is a prime, the graphs in (7.4.3) are precisely those of [LPS2]. It is, of course, also possible to extend the arguments in [LPS2] to the case of general N.

(iv) The computations of this section are made possible because the algebra of Hamiltonian quaternion has class number one (i.e., every ideal of the maximal order spanned over $\mathbb{Z}$ by $1, i, j$ and $\frac{1}{2}(1 + i + j + k)$ is principal). This implies that $\Gamma = G'(\mathbb{Z}[\frac{1}{p}])$ acts transitively on the tree T (cf. [Vi 1, p. 87]). We were able to find in Γ a congruence subgroup $\Gamma(2)$ which still acts transitively, but moreover also freely. By picking generators of $\Gamma(2)$ that take the base point x_0 (the standard lattice in $\mathbb{Q}_p \times \mathbb{Q}_p$) to all of its neighbors, we were able to identify the tree with the Cayley graph of $\Gamma(2)$. This is what makes $\Gamma(N)\backslash T$ a Cayley graph. It is possible to follow the same pattern for $p \equiv 3 \pmod 4$. In fact, the computations made in [GVDP, Ch. IX, 1] present the appropriate generators for $\Gamma(2)$. (This won't be a free group, rather a free product of infinite cyclic groups and cyclic groups of order 2. The Cayley graph of such a group is, however, still a tree!) This procedure was carried out by Sarnak [Sa 3].

The case of $p = 2$ needs some more modifications: Here we should replace the algebra by one which splits in $\mathbb{Q}_2$ and has class number one. $D = D(2, 13)$ will do. Next, one should find a congruence subgroup which acts simply transitively on the tree. P. Chiu [Chi] carried out these computations.

(v) We don't know whether this is a general fact that whenever $\Gamma = \Gamma(1)$ of (7.3.1) acts transitively on the tree, there exists a (congruence) subgroup acting simply transitive on the tree. It is not difficult to see that there is a finite index subgroup acting fixed point free, but it is not clear if such a subgroup can be found, in general, that still acts transitively. This does not seem to be the case for a general (non-arithmetic) lattice, and if it is true (as some examples suggest) for arithmetic lattices it might have an interesting arithmetic explanation. (The fact that $\Gamma(2)$ works for the Hamiltonian quaternions and 2 is the only finite prime at which it ramifies might be suggestive.)

We end this section by bringing explicit constructions of k-regular Ramanujan graphs for $k = q + 1$ when q is an arbitrary prime power. These graphs were constructed by Morgenstern [Mor1] by applying an analogue procedure to the one described here but replacing $\mathbb{Q}$ by a global field of characteristic $p > 0$. The analogue of Ramanujan conjecture is a theorem of Drinfeld [Dr2].

Theorem 7.4.5. *Let q be an odd prime power, ε a non-square in $\mathbb{F}_q$. Let $g(x) \in \mathbb{F}_q[x]$ be irreducible of even degree d, and $\mathbb{F}_{q^d}$ is represented as $\mathbb{F}_q[x]/g(x)\mathbb{F}_q[x]$. Let $\underline{\mathbf{i}} \in \mathbb{F}_{q^d}$ be such that $\underline{\mathbf{i}}^2 = \varepsilon$, and*

$$\begin{pmatrix} 1 & \gamma_k - \delta_k \underline{\mathbf{i}} \\ (\gamma_k + \delta_k \underline{\mathbf{i}})(x-1) & 1 \end{pmatrix} \qquad k = 1, \ldots, q+1 \qquad (*)$$

where $\gamma_k, \delta_k \in \mathbb{F}_q$, are all the $q + 1$ solutions in $\mathbb{F}_q$ for $\delta_k^2 \varepsilon - \gamma_k^2 = 1$.

(a) *If $\left(\frac{x}{g(x)}\right) = 1$.*

Let Ω_g be the Cayley graph of $PSL_2(\mathbb{F}_{q^d})$ with respect to the generators $(*)$. Then Ω_g is a $(q+1)$-regular Ramanujan graph, which is not bi-partite.

(b) *If $\left(\frac{x}{g(x)}\right) = -1$.*

Let Ω_g be the Cayley graph of $PGL_2(\mathbb{F}_{q^d})$ with respect to the generators $(*)$. Then Ω_g is a $(q+1)$-regular bi-partite Ramanujan graph.

Theorem 7.4.6. *Let q be a power of 2 and $f(x) = x^2 + x + \varepsilon$ an irreducible polynomial in $\mathbb{F}_q[x]$. Let $g(x) \in \mathbb{F}_q[x]$ be irreducible of even degree d, and $\mathbb{F}_{q^d}$ is represented as $\mathbb{F}_q[x]/g(x)\mathbb{F}_q[x]$. Let $\underline{\mathbf{i}} \in F_{q^d}$ be a root of $f(x)$, and*

$$\begin{pmatrix} 1 & \gamma_k + \delta_k \underline{\mathbf{i}} \\ (\gamma_k + \delta_k \underline{\mathbf{i}} + \delta_k)x & 1 \end{pmatrix} \qquad k = 1, \ldots, q+1 \qquad (**)$$

$\gamma_k, \delta_k \in \mathbb{F}_q$, are all the $q+1$ solutions in $\mathbb{F}_q$ for $\gamma_k^2 + \gamma_k\delta_k + \delta_k^2\varepsilon = 1$. Let Ω_g be the Cayley graph of $PSL_2(\mathbb{F}_{q^d})$ with respect to the generators (**). Then Ω_g is a $(q+1)$-regular Ramanujan graph, which is not bi-partite.

The graphs presented in Theorems (7.4.5.–7.4.6.) also enjoy the other properties (of girth, of chromatic number, etc.) as in Theorem (7.3.12).

7.5 Notes

1. The first Ramanujan graphs in the literature seem to appear implicitly in the paper of Ihara [Ih]. A matrix A is defined there which is indeed the adjacency matrix of $\Gamma(1)\backslash T$. When varying the algebras (instead of varying the congruence subgroups) we get infinitely many Ramanujan graphs since the class numbers of the quaternion algebras go to ∞. These graphs seem to be very interesting from an arithmetic point of view, but it is probably impossible to compute them explicitly. They also do not seem to have the other properties described in (7.3.12). On the other hand their zeta function (for the notion of zeta function of a graph, see 4.5 and [Ha]) is, up to trivial factors, a zeta function of a suitable curve over a finite field. Moreover, the statement that these graphs are Ramanujan follows at once from the Riemann hypothesis for curves over finite fields proved by Weil. See [Ih] and [Ha] for more details. It will be very interesting to relate some curves to $\Gamma(N) \setminus T$ and to study their properties.
2. The Ramanujan graphs constructed in this chapter give the best known expanders. They, therefore, improve all the various networks using expanders (see 1.1) and in particular give the best explicit super-concentrators known. But, they still fall short from those known to exist by counting arguments (and those are also not necessarily the best possible). It might be that indeed they are expanders as good as the random ones but we do not know to prove it. Pippenger (unpublished) considered these questions in detail. It is, however, still a challenge to come up with new methods to construct expanders and maybe some whose mathematics is less involved than the one presented in this book.
3. An interesting combinatorial invariant of a graph which is also related to eigenvalues is the number of spanning trees of X, denoted $K(X)$. By the Kirchhoff formula (see, for example, [B1, Theorem 8, p. 40]) $K(X) = \frac{1}{n}\prod_{i=1}^{n-1} \mu_i$, where $\mu_1, \ldots, \mu_{n-1}$ are the nonzero eigenvalues of the Laplacian of X. Let
$$C(X) = K(X)^{\frac{1}{n}}, \qquad \gamma(k) = \lim_{n\to\infty} \sup C(X_{n,k})$$
and
$$\beta(k) = \lim_{n\to\infty} \inf C(X_{n,k})$$

where the sup and inf are over the k-regular graphs with n vertices. It was shown by Sarnak [Sa3] that for $k \geq 3$, $\gamma(k) = \frac{(k-1)^{k-1}}{(k(k-2))^{\frac{k}{2}-1}}$ and that the Ramanujan graphs discussed in this chapter have $\gamma(k)$ as their limit. So, this is another extremal property they have.

It is not directly related, but we cannot resist the temptation to mention a result of Alon [A3]: $\beta(k) \geq k - 0(k\frac{(\log\log k)^2}{\log k})$. This purely combinatorial result has (as observed by Sarnak [Sa 3]) non-trivial arithmetic corollaries! It gives an estimation of the class numbers of some interesting function fields (see also [Ha]).

What we have been doing in this chapter were applications of number theory to combinatorics. This last result is a first step in the opposite direction! The paper of Pays [Pa] also has a similar flavor.

4. It was proved by J. Friedman [Fr1] that a random k-regular graph is not far from being Ramanujan. In the notation above he proved that for a fixed even integer k, most k-regular graphs on $n \to \infty$ vertices have $\lambda \leq 2\sqrt{k-1} + 2\log k + c$ for every eigenvalue λ which is not k, and where c is a constant. This was conjectured by Alon [A1], and a weaker statement was proved before by Broder and Shamir [BS].
5. S. Mozes [Mz] used the Ramanujan graphs constructed in this chapter to construct «tilling systems» having zero entropy and mixing of all orders (with explicit quantitative estimates of the rate of mixing).

8 Some More Discrete Mathematics

8.0 Introduction

This chapter is devoted to several results on graphs or groups which are related to the topics discussed above. We begin with an application to finite simple groups: Every finite simple non-abelian group G has a set S of at most seven generators with respect to which every element of G can be written as a word of length $O(\log |G|)$ with elements from $S \cup S^{-1}$. This theorem is proved in Babai-Kantor-Lubotzky [BKL] by finite group theoretic methods but with «unnatural» generators. It turns out that Property (T) and Selberg's Theorem give similar results in special cases with «natural» generators. The use of these deep theorems seems at this point to be unavoidable.

In the second section, we present a method from Diaconis-Shahshahani [DSh] to evaluate the eigenvalues of the adjacency matrix of a Cayley graph of a group G with respect to a full conjugacy class as a set of generators. In [DSh] this is done for the symmetric group, but here, following a suggestion of M. Ben-Or, we apply it for $SL_2(q)$. The method works easily for any group G once the character table is given. In $SL_2(q)$ we see that with respect to some conjugacy classes the graphs are Ramanujan graphs. Similarly, some of the graphs presented by Chung in [C2] are Ramanujan graphs. In all these cases the Ramanujan graphs are of unbounded degree, so they cannot be considered as a solution to the problem of expanders discussed in Chapter 1.

The same remark applies to some more graphs – the generalized N-gons – which are discussed in 8.3. These graphs are very interesting, even though by a well-known result of Feit and Higman [FH] their existence is limited. Those that do exist give rise to Ramanujan graphs! An interesting observation due to P. de la Harpe is that the well-known Paley graphs are also Ramanujan graphs – but again of unbounded degree.

A different interesting direction is described in 8.4: Here following Morgenstern ([Mor2], [Mor3]) we present «Ramanujan diagrams». Diagrams are graphs with weights which can be infinite but with finite total volume. The theory of expanders, Laplacian, eigenvalues, etc. can be extended to them. Moreover, in analogue to what was described in Chapter 7, Ramanujan diagrams can be built up from lattices in $\mathrm{PGL}_2(\mathbb{F}((\frac{1}{t})))$. But, this time noncompact lattices are used instead of cocompact. This notion of Ramanujan diagrams is not merely of theoretical interest; quite surprisingly they led to better bounded concentrators (see 1.1.7). So the excitements of Ramanujan conjecture are not yet over. . . .

8.1 The diameter of finite simple groups

It was noticed by Alon and Milman in [AM] that some of the results described in the previous chapters have interesting corollaries in group theory. More specifically:

Proposition 8.1.1. *Fix $n \in \mathbb{N}$ with $n \geq 2$ and let $F = \mathbb{F}_p$ be the finite field with p elements. Let A_n and B_n be the matrices in $SL_n(F)$ acting on the standard basis $\{e_1, \ldots, e_n\}$ of F^n as follows: $A_n(e_1) = e_1 + e_2$ and $A_n(e_i) = e_i$ for $i \neq 1$, while $B_n(e_i) = e_{i+1}$ for $1 \leq i \leq n-1$ and $B_n(e_n) = (-1)^{n-1}e_1$. Then there exists a constant C_n independent of p such that every element of $SL_n(p)$ can be written as a word of length at most $C_n \log p$ using elements from the set $S = \{A_n^{\pm 1}, B_n^{\pm 1}\}$.*

Proof. The claim of the proposition simply says that the diameter of the Cayley graph $X(SL_n(p); S)$ is $\leq C_n \log(p)$.

This is indeed the case: For $n \geq 3$ (and n is fixed!) these graphs form a family of expanders (3.3.2). It is easy to prove that a family of expanders has diameter $O(\log |X|)$. (An indirect way to see it is to use (4.2.4) and (7.3.11-ii).) The same conclusion holds for $n = 2$ by (4.4.2). □

It is interesting to note that the case $n \geq 3$ is a corollary of Kazhdan Property (T) while $n = 2$ is a corollary of Selberg's theorem. We don't know an elementary proof for either case, nor do we know an effective way to present an element of the group as a short word using A_n and B_n. The case $n = 2$ is especially challenging.

Problem 8.1.2 *Does there exist a polynomial time algorithm (polynomial in $\log p$) which expresses an element of $SL_2(p)$ (say $\begin{pmatrix} 1 & \frac{p-1}{2} \\ 0 & 1 \end{pmatrix}$) as a short word (say less than $1000 \log p$) in $A = \begin{pmatrix} 1 & 1 \\ 0 & 1 \end{pmatrix}$ and $B = \begin{pmatrix} 0 & 1 \\ -1 & 0 \end{pmatrix}$?*

This problem can be thought of as a non-commutative analogue of the discrete logarithm problem: there, one wants to express a given element x of $\mathbb{F}_p^* = GL_1(\mathbb{F}_p)$ as a power of a primitive element g, i.e., to find «$\log_g(x)$». In that problem, however, $\log_g(x)$ is unique $(\mathrm{mod}(p-1))$, so there are no «long» or «short» expressions.

Proposition (8.1.1) suggests another interesting question, which is how C_n varies with n. Since the order of $SL_n(p)$ is approximately p^{n^2-1} and the diameter of a group G is at least $\log(|G|)$ (see the remarks at the end of 7.3), the best we can hope for is $C_n = O(n^2)$. We do not know if this is the case.

Problem 8.1.3. *Does there exist a constant C such that every element of $SL_n(p)$ can be written as a word of length $\leq Cn^2 \log p$ in $A_n^{\pm 1}$ and $B_n^{\pm 1}$?*

What we do know is that using a different set of generators, this is indeed possible. This follows from the following theorem of Babai-Kantor-Lubotzky [BKL], whose proof there uses some «unnatural» generators for $SL_n(p)$ but ones for which the proof is elementary (i.e., without Kazhdan or Selburg) and the generators are «effective» (i.e., an algorithm for the short presentation is given).

Theorem 8.1.4. *There exists a constant C such that every finite simple non-abelian group G has a set S of seven generators such that* $\mathrm{diam}(X(G; S)) \leq C \log(|G|)$, *i.e., every element of G can be written as a product of elements of $S \cup S^{-1}$ of length $\leq C \log(|G|)$.*

Before giving a (partial) proof for this theorem, we compare it with:

Proposition 8.1.5. *Let $G_p = C_p$ the cyclic group of order p and S a set of k elements generating G_p. Then* diam $X(G_p; S) \geq \frac{1}{2}p^{1/k}$. *In particular, the conclusion of Theorem 8.1.4 does not hold for the family G_p.*

Proof. If diameter $X(G_p; S) = d$ then, since G_p is commutative, every $g \in G_p$ can be written as $g = s_1^{\alpha_1} s_2^{\alpha_2} \cdots s_k^{\alpha_k}$ where $S = \{s_1, \dots, s_k\}$ and $|\alpha_i| \leq d$. Hence $(2d+1)^k \geq |G_p|$. □

From the proof one sees that what plays the main role is the «polynomial growth» of abelian groups. The proposition can be extended to families of finite nilpotent groups of bounded nilpotency class. (Compare Bass [Bss1], Gromov [Gv] and [AB].) Families of solvable groups might have logarithmic diameter. For example, let $V_n = C_2 \times \ldots \times C_2$ the elementary abelian 2-group on n generators. The cyclic group $C_n = \langle g \rangle$ acts on it by a cyclic shift. Let $G(n)$ be the semi-direct product $V_n \rtimes C_n$. It is not difficult to check that $S = \{(1, 0, \dots, 0), g\}$ generates $G(n)$ in $O(\log(|G(n)|)) = O(\log(2^n \cdot n)) = O(n)$ steps. This happens, even though the graphs $X(G(n), S)$ are not expanders by Proposition (3.3.7). (The groups $G(n)$ are all quotients of the metabelian free group on two generators which is amenable!)

It would be interesting to formulate a general result which will describe conditions on families of groups to have (or not have) logarithmic diameter. This property depends also on the choice of generators as the following important example shows:

Proposition 8.1.6. *Let S_n be the symmetric group of all permutations of $\{1, \dots, n\}$ letters.*

(i) *Let $\tau = (1, 2)$ and $\sigma = (1, \dots, n)$. Then there exist constants C_1, C_2 such that:*

$$C_1 n^2 \leq \text{ diam } X(S_n, \{\tau, \sigma\}) \leq C_2 n^2.$$

(ii) *There exists a two-generator set Σ of S_n for which* diam $X(S_n, \Sigma) = O(n \log n) = O(\log |S_n|)$.

Proof. (i) Every permutation is a product of at most $2n$ transpositions of type $(1, i)$, $2 \leq i \leq n$. We will show that $(1, i)$ can be written as a word of length $O(n)$ in τ and σ. Indeed: let $\psi = \tau\sigma = (2, 3, \dots, n)$, then $(1, i) = \psi^{i-2}(1, 2)\psi^{-(i-2)}$. This proves the right-hand side inequality with $C_2 = 4$. To prove the lower bound: Given a permutation γ and a triple $i < j < k$ of different integers between 1 and n, (i, j, k) is said to be in a good position with respect to γ if either $\gamma(i) < \gamma(j) < \gamma(k)$ or $\gamma(j) < \gamma(k) < \gamma(i)$ or $\gamma(k) < \gamma(i) < \gamma(j)$. It is easy to check that (i, j, k) is in a good position with respect to γ if and only if it is in a good position with respect to $\sigma\gamma$. If (i, j, k) is in a bad position with respect to γ then it might be in a good position with respect to $\tau\gamma$ only if $\{\gamma(i), \gamma(j), \gamma(k)\} = \{1, 2, \ell\}$ for some ℓ. So multiplication by τ can «mend» at most n triples. Now, for $\gamma = (1, n)(2, n-1)\cdots(i, n+1-i)\cdots([\frac{n}{2}], n-[\frac{n}{2}]+1)$ all the $\binom{n}{3}$ triples are bad, and so any word expressing γ by τ and σ should include τ at least $\frac{n^2}{6}$ times.

(ii) For simplicity of notation we assume n is even, and we will present three generators and not two as promised (for a full proof see [BKL]).

Think of S_n as the group of permutations of $\{\infty\} \cup \mathbb{Z}/(n-1)\mathbb{Z} = \{0, 1, \ldots, n-2, \infty\}$. Consider the two permutations $\beta_0 : x \to 2x(\mathrm{mod}(n-1))$, and $\beta_1 : x \mapsto 2x + 1(\mathrm{mod}(n-1))$. So, both fix ∞. Any element $t \in \mathbb{Z}/(n-1)\mathbb{Z}$ can be written $t = \sum_{i=0}^{m} a_i 2^i = (\cdots(a_m \cdot 2 + a_{m-1})2 + \cdots)2 + a_0$ where $m = [\log_2 n]$ and each $a_i \in \{0, 1\}$. (The second equality is «Horner's rule».) Thus $t = w(0)$ where $w = \beta_{a_0} \ldots \beta_{a_{m-1}} \beta_{a_m}$. Now as in (i), if we take $\alpha = (0, \infty)$ to prove that $\{\alpha, \beta_0, \beta_1\}$ generate S_n in $O(n \log n)$, it suffices to prove that for every $t \in \{1, \ldots, n-2\}$ the transposition (∞, t) can be written in $O(\log n)$ steps. Indeed, $(\infty, t) = w(0, \infty)w^{-1}$, and we are done. □

It is a beautiful result of J. Dixon [Dix] that «almost» every pair of elements of S_n generates either S_n or the alternating group A_n. It is an interesting problem to determine what is the diameter of S_n with respect to such a random pair of generators.

We can now go back to the proof of (8.1.4). By the recent classification of the finite simple groups (cf. [Go]) we have to deal with A_n, groups of Lie type and 26 sporadic groups.

The case of A_n was essentially covered once we did S_n (the details are left to the reader; see [BKL]). There are only finitely many sporadic groups, and each of them can be generated by two elements – so they can be ignored. For the groups of Lie type we will present the proof only for the family $PSL_n(p)$ (or $SL_n(p)$) for p a prime. The general case can be treated in a similar way, but technically it is much more difficult and requires some knowledge of the structure of groups of Lie type, which we prefer not to assume. The rank one case also needs some special consideration – see [BKL].

Proposition 8.1.7. *There exists a constant C such that $SL_n(p)$ has a set S of three generators with* $\mathrm{diam}X(SL_n(p); S) \leq Cn^2 \log p$.

Proof. For $SL_2(p)$ we can give 2 generators $t = \begin{pmatrix} 1 & 1 \\ 0 & 1 \end{pmatrix}$ and $\tau = \begin{pmatrix} 0 & 1 \\ -1 & 0 \end{pmatrix}$ which generate it in $O(\log p)$ steps (8.1.1). By embedding $SL_2(p)$ at the left upper corner of $SL_n(p)$, they can be considered as elements of $SL_n(p)$. We add a third generator $\sigma = B_n$, where B_n is the same B_n from Proposition 8.1.1.

By the Bruhat decomposition (see [Car]), $G = BWB$ where B is the group of upper triangular matrices in G and W is the Weyl group generated by τ and σ. Indeed $W \simeq S_n$ and τ and σ correspond to τ and σ in Proposition (8.1.6.(i)), so they generate W in $O(n^2)$ steps, which for our purpose is fine!

Now $B = TU$ where T is the group of diagonal matrices of determinant one, and U the unipotent radical of B. We argue separately for T and U.

Let $\gamma \in T$, $\gamma = \begin{pmatrix} \lambda_1 & & & \\ & \lambda_2 & & \\ & & \cdots & \\ & & & \lambda_n \end{pmatrix}$ with $\prod_{i=1}^{n} \lambda_i = 1$. Then:

$$\gamma = \begin{pmatrix} \lambda_1 & & & & & \\ & \lambda_1^{-1} & & & & \\ & & 1 & & & \\ & & & \cdots & & \\ & & & & 1 & \\ & & & & & 1 \end{pmatrix} \cdot \begin{pmatrix} 1 & & & & & \\ & \lambda_1\lambda_2 & & & & \\ & & (\lambda_1\lambda_2)^{-1} & & & \\ & & & 1 & & \\ & & & & \cdots & \\ & & & & & 1 \end{pmatrix} \cdot$$

$$\cdots \cdot \begin{pmatrix} 1 & & & & \\ & 1 & & & \\ & & \cdots & & \\ & & & \prod_{i=1}^{n-1} \lambda_i & \\ & & & & (\prod_{i=1}^{n-1} \lambda_i)^{-1} \end{pmatrix}$$

Hence

$$\begin{aligned} \gamma &\in SL_2(p) \cdot \sigma SL_2(p)\sigma^{-1} \cdot \sigma^2 SL_2(p)\sigma^{-2} \ldots \sigma^{n-2} SL_2(p)\sigma^{-(n-2)} \\ &= SL_2(p) \cdot \sigma \cdot SL_2(p) \cdot \sigma \cdot SL_2(p) \cdot \sigma \ldots \sigma \cdot SL_2(p)\sigma^{-(n-2)}. \end{aligned}$$

The «price» of γ is therefore at most $O(n \log p)$.

Now, let $\gamma \in U$. It is known that U can be written as a product of the elementary subgroups X_{ij} in any order we wish. Here X_{ij} is the subgroup $\{1 + aE_{ij} \mid a \in \mathbb{F}_p\}$ where E_{ij} is the matrix with 1 at entry (i, j) and 0 elsewhere. Then $U = \prod_{i=1}^{n} \prod_{k=0}^{n-i-1} X_{i,i+1+k}$. Thus

$$\begin{aligned} U &= Y(\sigma Y \sigma^{-1})(\sigma^2 Y \sigma^{-2}) \ldots (\sigma^{n-1} Y \sigma^{-(n-1)}) \\ &= Y\sigma Y \sigma Y \sigma \ldots \sigma Y \sigma^{-(n-1)} \end{aligned}$$

where

$$\begin{aligned} Y &= X_{1,2} \cdot (\psi X_{1,2} \psi^{-1}) \cdot (\psi^2 X_{1,2} \psi^{-2}) \ldots \psi^{n-2} X_{1,2} \psi^{-(n-2)} \\ &= X_{1,2} \cdot \psi \cdot X_{1,2} \cdot \psi \cdot X_{1,2} \cdot \psi \ldots \psi \cdot X_{1,2} \psi^{-(n-2)} \end{aligned}$$

where $\psi = \tau\sigma$ is the element of W corresponding to $(2, 3, \ldots, n)$ in S_n. All this (and especially the wonderful cancellations occurring in the above products) shows that the price of $\gamma \in U$ is at most $O(n^2 \log p)$ (since $X_{1,2} \subseteq SL_2(p)$). The proposition is therefore proven. $\square$

Remarks 8.1.8.

(i) In the proof we used Selberg's Theorem to get an «efficient» two generators set for $SL_2(p)$. This is an overkill. It is possible to find by elementary means such a set (see [BKL]) but it is not as «nice» as $t = \begin{pmatrix} 1 & 1 \\ 0 & 1 \end{pmatrix}$ and $\tau = \begin{pmatrix} 0 & 1 \\ -1 & 0 \end{pmatrix}$.

(ii) Note that Proposition 8.1.7 *does not* answer Problem 8.1.3 since it gives a different set of generators.

(iii) Unlike the case of S_n (8.1.6), we do not know any «bad» choice of generators for $SL_n(p)$. Again, the case of $SL_2(p)$ is of special interest.

8.2 Characters and eigenvalues of finite groups

In this section, let G be a finite group, S a symmetric subset of G, and $X = X(G; S)$ the Cayley graph of G with respect to S. We will show, following Diaconis-Shahshahani [DSh], how the representation theory (more specifically, the character table) of G can be useful in evaluating eigenvalues of the adjacency matrix δ. This method is especially useful when S is a union of conjugacy classes. Following a suggestion of M. Ben-Or we apply it to the group $SL_2(q)$ to get Ramanujan graphs. This is an easy way to get Ramanujan graphs, but the graphs so obtained are of unbounded degree and so are not as useful as those in Chapter 7. Still, they are interesting and the point of view is also relevant for the understanding of the graphs of Chung [C2] to be discussed afterwards.

A short excellent reference for representation theory of finite groups is Serre's book [S3]. Interesting applications of representation theory can be found in Diaconis [D].

The basic connection between characters and eigenvalues is the following (cf. [B] or [DSh]):

Let δ be as before; it acts on $L^2(X)$. The latter can be identified with the group algebra $\mathbb{C}[G]$, when the convolution product in $L^2(X)$ is replaced by the ordinary product of the group algebra. The matrix δ can be thought of as an element of $\mathbb{C}[G]$, i.e., $\delta = \sum_{s \in S} s$. Then its action on $\mathbb{C}[G]$ is simply by multiplication from the right.

As is well known, $\mathbb{C}[G]$ as a right G-module is isomorphic to $\bigoplus_{i=1}^{r} d_i V_i$ when $(V_1, \rho_1), \ldots, (V_r, \rho_r)$ are the r different irreducible representations of G and $d_i =$

dim V_i. Moreover, as an algebra $\mathbb{C}[G] \simeq \bigoplus_{i=1}^{r} M_{d_i}(\mathbb{C})$. The isomorphism is given by: $\alpha \longmapsto (\rho_1(\alpha), \dots, \rho_r(\alpha))$ where $\alpha \in \mathbb{C}[G]$ and ρ_i is extended in a linear way to a homomorphism $\rho_i : \mathbb{C}[G] \longrightarrow \mathrm{End}(V_i) = M_{d_i}(\mathbb{C})$. In particular, the multiplication by δ from the right is indeed the direct sum of applying $\rho_i(\delta)$ as multiplication from the right on $M_{d_i}(\mathbb{C})$.

Now, if S is a union of conjugacy classes, then it is not difficult to check (see also [S3, p. 50]) that δ is in the center of $\mathbb{C}[G]$. From Schur's lemma it follows that for every i, $\rho_i(\delta)$ is a central (i.e., a scalar matrix $\mu_i I$) in $M_{d_i}(\mathbb{C})$. So, its action on $M_{d_i}(\mathbb{C})$ is by multiplication by that scalar μ_i. The eigenvalues of the adjacency matrix δ are, therefore, $\mu_1, \dots, \mu_r$ with multiplicities $d_1^2, \dots, d_r^2$, respectively. It remains to compute the $\mu_i - s$. This is fairly easy if one knows the character table of G. The character table tells us what is $\chi_i(s) = Tr(\rho_i(s))$ for every $i = 1, \dots, r$ and $s \in S$. So: $Tr(\rho_i(\delta)) = Tr(\rho_i(\sum_{s \in S} s)) = \sum_{s \in S} Tr(\rho_i(s)) = \sum_{s \in S} \chi_i(s)$. At the same time, by the discussion above $\rho_i(\delta) = \mu_i I$ and hence $\mu_i = \frac{Tr(\rho_i(\delta))}{d_i} = \frac{1}{d_i} \sum_{s \in S} \chi_i(s)$.

If S is precisely one conjugacy class then $\chi_i(s)$ are all the same for $s \in S$, and then $\mu_i = \frac{|S|}{d_i} \chi_i(s)$.

To summarize:

Theorem 8.2.1. *Let G be a finite group and S a symmetric union of conjugacy classes of G, $S = \bigcup_{j=1}^{k} C_j$. Let X be the Cayley graph of G with respect to S, and δ the adjacency matrix of X. Let $\rho_1, \dots, \rho_r$ be the irreducible representations of G and $\chi_1, \dots, \chi_r$ the corresponding characters, i.e., $\chi_i(g) = Tr(\rho_i(g))$ for $g \in G$ ($r =$ the number of conjugacy classes of G).*

Then the eigenvalues of δ are $\mu_1, \dots, \mu_r$ with multiplicities $d_i^2 = (\dim \rho_i)^2 = \chi_i^2(e)$ where e is the identity element of G and

$$\mu_i = \frac{1}{d_i} \sum_{s \in S} \chi_i(s) = \frac{1}{d_i} \sum_{j=1}^{k} |C_j| \chi_i(C_j)$$

for $i = 1, \dots, r$. If S is a unique conjugacy class, then $\mu_i = \frac{|S|}{d_i} \chi_i(S) = |S| \frac{\chi_i(S)}{\chi_i(e)}$, when $\chi(C)$ means $\chi(c)$ where c is one of the elements of the conjugacy class C.

The theorem gives us a method to compute *all* the eigenvalues of Cayley graphs of groups provided the character table is known and the set of generators we are using is a union of conjugacy classes. The union should be symmetric, i.e., if the conjugacy class of s is in S then so is the conjugacy class of s^{-1}. (Of course, in many cases s is conjugate to s^{-1}.) Otherwise we get the eigenvalues of a directed graph and the eigenvalues are not necessarily real.

For example, let $G = SL_2(q)$ where $q = p^n$, p an odd prime. The following theorem quoted from [Dor, Theorem 38.1] describes, in full, the character table of G. As an immediate corollary afterward we will deduce that some graphs are Ramanujan.

Theorem 8.2.2. *Let F be the finite field with $q = p^n$ elements, p an odd prime, and let ν be a generator of the cyclic group $F^* = F - \{0\}$. Denote*

$$1 = \begin{pmatrix} 1 & 0 \\ 0 & 1 \end{pmatrix}, \quad z = \begin{pmatrix} -1 & 0 \\ 0 & -1 \end{pmatrix},$$

$$c = \begin{pmatrix} 1 & 0 \\ 1 & 1 \end{pmatrix}, \quad d = \begin{pmatrix} 1 & 0 \\ \nu & 1 \end{pmatrix}, \quad a = \begin{pmatrix} \nu & 0 \\ 0 & \nu^{-1} \end{pmatrix}$$

in $G = SL(2, F)$. G contains an element b of order $q + 1$, and we fix such one element.

For any $x \in G$, let (x) denote the conjugacy class of G containing x. Then G has exactly $q + 4$ conjugacy classes (1), (z), (c), (d), (zc), (zd), (a), (a^2), $\dots$, $(a^{(q-3)/2})$, (b), (b^2), $\dots$, $(b^{(q-1)/2})$, satisfying

x	1	z	c	d	zc	zd	a^ℓ	b^m
$\lvert(x)\rvert$	1	1	$\frac{1}{2}(q^2-1)$	$\frac{1}{2}(q^2-1)$	$\frac{1}{2}(q^2-1)$	$\frac{1}{2}(q^2-1)$	$q(q+1)$	$q(q-1)$

for $1 \le \ell \le (q-3)/2, 1 \le m \le (q-1)/2$.

Denote $\varepsilon = (-1)^{(q-1)/2}$. Let $\rho \in \mathbb{C}$ be a primitive $(q-1)$-th root of 1, $\sigma \in \mathbb{C}$ a primitive $(q+1)$-th root of 1. Then the complex character table of G is

	1	z	c	d	a^ℓ	b^m
1_G	1	1	1	1	1	1
ψ	q	q	0	0	1	-1
χ_i	$q+1$	$(-1)^i(q+1)$	1	1	$\rho^{i\ell}+\rho^{-i\ell}$	0
θ_j	$q-1$	$(-1)^j(q-1)$	-1	-1	0	$-(\sigma^{jm}+\sigma^{-jm})$
ξ_1	$\frac{1}{2}(q+1)$	$\frac{1}{2}\varepsilon(q+1)$	$\frac{1}{2}(1+\sqrt{\varepsilon q})$	$\frac{1}{2}(1-\sqrt{\varepsilon q})$	$(-1)^\ell$	0
ξ_2	$\frac{1}{2}(q+1)$	$\frac{1}{2}\varepsilon(q+1)$	$\frac{1}{2}(1-\sqrt{\varepsilon q})$	$\frac{1}{2}(1+\sqrt{\varepsilon q})$	$(-1)^\ell$	0
η_1	$\frac{1}{2}(q-1)$	$-\frac{1}{2}\varepsilon(q-1)$	$\frac{1}{2}(-1+\sqrt{\varepsilon q})$	$\frac{1}{2}(-1-\sqrt{\varepsilon q})$	0	$(-1)^{m+1}$
η_2	$\frac{1}{2}(q-1)$	$-\frac{1}{2}\varepsilon(q-1)$	$\frac{1}{2}(-1-\sqrt{\varepsilon q})$	$\frac{1}{2}(-1+\sqrt{\varepsilon q})$	0	$(-1)^{m+1}$

for $1 \le i \le (q-3)/2$, $1 \le j \le (q-1)/2$, $1 \le \ell \le (q-3)/2$, $1 \le m \le (q-1)/2$.

(The columns for the classes (zc) and (zd) are missing in this table. These values are obtained from the relations

$$\chi(zc) = \frac{\chi(z)}{\chi(1)}\chi(c), \qquad \chi(zd) = \frac{\chi(z)}{\chi(1)}\chi(d),$$

for all irreducible characters χ of G.)

Corollary 8.2.3. *(i) Let* $G = SL_2(q)$*, assume* $q \equiv 1 (\mathrm{mod}\, 4)$ *and let* S *be the conjugacy class of* $\begin{pmatrix} 1 & 0 \\ 1 & 1 \end{pmatrix}$ *in* G*. Then* $|S| = \frac{1}{2}(q^2 - 1)$ *and the eigenvalues of* $X(G;S)$ *are:*

eigenvalues	$\frac{1}{2}(q^2-1)$	0	$\frac{1}{2}(q-1)$	$-\frac{1}{2}(q+1)$	$\frac{1}{2}(1+\sqrt{q})(q-1)$
multiplicities	1	q^2	$\frac{1}{2}(q-3)(q+1)^2$	$\frac{1}{2}(q-1)^3$	$\frac{1}{4}(q+1)^2$

eigenvalues	$\frac{1}{2}(1-\sqrt{q})(q-1)$	$\frac{1}{2}(-1+\sqrt{q})(q+1)$	$\frac{1}{2}(-1-\sqrt{q})(q+1)$
multiplicities	$\frac{1}{4}(q+1)^2$	$\frac{1}{4}(q-1)^2$	$\frac{1}{4}(q-1)^2$

Since X *is a* k*-regular graph for* $k = |S| = \frac{1}{2}(q^2 - 1)$ *and there are eigenvalues of size* $\approx q^{\frac{3}{2}}$*,* X *is not a Ramanujan graph.*

(ii) Let $G = SL_2(q)$ *and assume* $q \equiv 3\ (\mathrm{mod}\, 4)$*. Let* S *be the union of the conjugacy class* C_1 *of* $\begin{pmatrix} 1 & 0 \\ 1 & 1 \end{pmatrix}$ *and* C_{-1} *of* $\begin{pmatrix} 1 & 0 \\ -1 & 1 \end{pmatrix}$*. Then* $|S| = q^2 - 1$ *and the eigenvalues of* $X = X(G;S)$ *are:*

eigenvalues	(q^2-1)	0	$q-1$	$-(q+1)$
multiplicities	1	q^2	$\frac{1}{2}(q-2)(q+1)^2$	$\frac{q}{2}(q-1)^2$.

Since X *is a* k*-regular graph for* $k = |S| = q^2 - 1$ *and* $|q+1| \leq 2\sqrt{q^2-1}$*,* X *is a Ramanujan graph!*

(iii) Let $G = SL_2(q)$ *and* S *the full conjugacy class of* $\begin{pmatrix} \nu & 0 \\ 0 & \nu^{-1} \end{pmatrix}$ *where* ν *is as in Theorem 8.2.2, a generator for the group* $\mathbb{F}_q^*$*. Then* $|S| = q(q+1)$ *and the eigenvalues of* $X(G;S)$ *are:*

eigenvalues	$q(q+1)$	$q+1$	0	$-2q$
multiplicities	1	q^2	$\frac{1}{2}q(q-1)^2$	$\frac{1}{2}(q+1)^2$

eigenvalues	For $i = 1, \ldots, (q-3)/2 : (\rho^i + \rho^{-i})q$
multiplicities	For each $i : (q+1)^2$

where ρ *is a primitive* $(q-1)$*-th root of* 1*. Since* X *is* k*-regular for* $k = q(q+1)$ *and* $2q \leq 2\sqrt{q(q+1)-1}$*,* X *is a Ramanujan graph!*

(iv) Let $G = SL_2(q)$*,* b *a fixed element of* $SL_2(q)$ *of order* $q+1$ *and* S *the conjugacy class of* b*. Then* $S = q(q-1)$ *and the eigenvalues of* $X = X(G;S)$ *are*

eigenvalues	$q(q-1)$	$-(q-1)$	0	$2q$
multiplicities	1	q^2	$\frac{1}{2}(q-2)(q+1)^2$	$\frac{1}{2}(q-1)^2$

eigenvalues	For $j = 1, \ldots, \frac{1}{2}(q-1) : -(\sigma^j + \sigma^{-j})q$
multiplicities	For each $j : (q-1)^2$

where σ *is a primitive* $(q+1)$*-th root of* 1*. Since* X *is* k*-regular for* $k = q(q-1)$ *and* $2q \geq 2\sqrt{q(q-1)-1}$*,* X *is not a Ramanujan graph.*

Proof. Everything follows from Theorem 8.2.1 by (tedious) calculations using the information in Theorem 8.2.2. We only remark that in cases (i), (iii) and (iv) the conjugacy class S is indeed symmetric: In case (i), we have $\eta \in \mathbb{F}_q$ such that $\eta^2 \equiv -1$, hence $\begin{pmatrix} \eta & 0 \\ 0 & \eta^{-1} \end{pmatrix}\begin{pmatrix} 1 & 0 \\ 1 & 1 \end{pmatrix}\begin{pmatrix} \eta^{-1} & 0 \\ 0 & \eta \end{pmatrix} = \begin{pmatrix} 1 & 0 \\ -1 & 1 \end{pmatrix}$ and so the inverse of $\begin{pmatrix} 1 & 0 \\ 1 & 1 \end{pmatrix}$ is conjugate to $\begin{pmatrix} 1 & 0 \\ 1 & 1 \end{pmatrix}$. (This is not the case in (ii), and this is why we had to take a second conjugacy class. Note that $\begin{pmatrix} 1 & 0 \\ \nu & 1 \end{pmatrix}$ is conjugate to $\begin{pmatrix} 1 & 0 \\ -1 & 1 \end{pmatrix}$ since -1 and ν both are not squares in $\mathbb{F}_q^*$.) In case (iii) it is clear that $\begin{pmatrix} \nu & 0 \\ 0 & \nu^{-1} \end{pmatrix}$ is conjugate to its inverse. Case (iv) (as well as (i) and (iii)) follows from the following fact: If $s \in G$ and for every character χ of G, $\chi(s)$ is real then s is conjugate to s^{-1}. To prove this fact, note that $\chi(s^{-1}) = \overline{\chi(s)}$ (cf. [S3, p. 10]) for every χ. This implies that s and s^{-1} are in the same conjugacy class. □

This last corollary illustrates that once we have the character table of a group, we can compute in great detail the eigenvalues of its Cayley graphs with respect to unions of conjugacy classes. This applies in particular to any subset of an abelian group, since with abelian groups the character table is easily computable and every single element is a conjugacy class. (Indeed, Theorem 8.2.1 for abelian groups is no more than standard Fourier analysis.) The tricky point here is to find a clever way to choose the generators to ensure good bounds on the eigenvalues. Such a method was given by F. Chung [C2]. First, a general easy result:

Proposition 8.2.4. *Let G be the cyclic group $\mathbb{Z}/n\mathbb{Z}$ of order n. Let $S = \{a_1, \ldots, a_k\}$ be a symmetric subset of G and $X = X(G;S)$. Then the eigenvalues of X are $\sum_{i=1}^{k} \theta^{a_i}$ where θ ranges over all n-th roots of 1.*

Proof. Either use (8.2.1), recalling that every irreducible representation is one-dimensional and given by $1 \longmapsto \theta$, or note that the vectors $(1, \theta, \theta^2, \ldots, \theta^{n-1})$ are eigenvectors. □

To make $|\sum_{i=1}^{k} \theta^{a_i}|$ small for $\theta \neq 1$, we have to specify the $a_i - s$ in a special way. Here is Chung's method:

Let p be a prime and $n = p^2 - 1$. Identify the cyclic group $\mathbb{Z}/n\mathbb{Z}$ with the multiplicative group of invertible elements of the finite field $K = \mathbb{F}_{p^2}$. This is done by adjoining to $\mathbb{F}_p$ a root w of an irreducible quadratic polynomial in $\mathbb{F}_p[x]$. Now, take an element $g \in K^*$ which generates K^* and consider the p elements $\bar{S}_0 = \{w, w+1, \ldots, w+p-1\}$. This is not a symmetric set. Let $\bar{S} = \bar{S}_0 \cup \bar{S}_0^{-1}$. It is not difficult to check that at most two elements in $\bar{S}_0$ have their inverses also in

$\bar{S}_0$, so $|\bar{S}| \geq 2p-2$. Clearly, $w+i$ can be expressed as g^{a_i} for some a_i, and these a_i, $0 \leq i \leq p-1$, form the subset S_0 of $\mathbb{Z}/n\mathbb{Z}$. Let $S = S_0 \cup (-S_0)$, $|S| \geq 2p-2$. The crucial point is the following theorem of N. Katz [Kat1]:

Theorem 8.2.5. *Let ψ be a non-trivial complex-valued multiplicative character defined on an extension field K of dimension t over a finite field F. Then for any $x \in K$ we have: $|\sum_{a\in F} \psi(x+a)| \leq (t-1)\sqrt{|F|}$.*

Applying the theorem to the circumstances above we deduce for θ an n-th root of 1 and $\psi(g) = \theta$:

$$|\sum_{x\in\mathbb{F}_p} \psi(\omega + x)| = |\sum_{i=0}^{p-1} \theta^{a_i}| \leq \sqrt{p}$$

and hence $|\sum_{i=0}^{p-1} \theta^{a_i} + \sum_{i=0}^{p-1} \theta^{-a_i}| \leq 2\sqrt{p}$.

Since the graph $X(\mathbb{Z}/n\mathbb{Z}; S)$ is regular of degree k (where $2p-2 \leq k \leq 2p$), we deduce

Theorem 8.2.6. *When S is chosen as above, the graphs $X(\mathbb{Z}/n\mathbb{Z}; S)$ are Ramanujan graphs.*

Moreover, F. Chung [C2] presents also other graphs which are similar to the above but are not Cayley graphs: Look at $G = \mathbb{Z}/n\mathbb{Z}$ and connect two elements $x, y \in G$ if $x + y = a_i$ for some $a_i \in S_0$ (S_0 and all notation as above). This is a symmetric relation! Such graphs are called «sum graphs».

Lemma 8.2.7. *The sum graph on $G = \mathbb{Z}/n\mathbb{Z}$, determined by the set $T = \{a_1, \ldots, a_k\}$ (i.e., $x, y \in G$ are adjacent iff $x + y \in T$), has eigenvalues $\pm|\sum_{i=1}^{k} \theta^{a_i}|$ where θ ranges over all n-th roots of 1.*

Proof. The eigenvectors are $(1, \theta, \theta^2, \ldots, \theta^{n-1}) + (\sum_{i=1}^{k} \theta^{a_i})/|\sum_{i=1}^{k} \theta^{a_i}| \cdot (1, \theta^{-1}, \ldots, \theta^{-(n-1)})$. □

Now Katz's Theorem 8.2.5 again implies that, if S_0 is as above, then the sum graph of $\mathbb{Z}/n\mathbb{Z}$ with respect to S_0 is a p-regular graph with all its eigenvalues are $\leq \sqrt{p}$. Hence

Theorem 8.2.8. *The sum graphs of $G = \mathbb{Z}/n\mathbb{Z}$ with respect to $S_0 = \{a_0, \ldots, a_{p-1}\}$ as above are Ramanujan graphs.*

We do not know a general theory to evaluate eigenvalues of sum graphs for non-abelian groups with respect to a subset T. It would be interesting to find an analogue of Theorem 8.2.1 for this case. (In the non-abelian case the graph comes with a natural orientation, and we may or may not ignore this orientation.)

Finally, we mentioned that in several cases the k-regular Ramanujan graphs presented in this chapter have eigenvalues λ with a bound on λ substantially better than $2\sqrt{k-1}$. This is not a contradiction to the Alon-Boppana Theorem (4.5.5) since we have such graphs only for finitely many values of n (given k). Thus, the graphs presented here do not form a family of expanders in the sense of Chapter 1. Still they show that for specific values (and hence in real world applications) one may expect graphs which have bounds on their eigenvalues which are substantially better than the Ramanujan bound. Similar remarks apply also for the graphs of the next section.

8.3 Some more Ramanujan graphs (of unbounded degrees)

In this section we mention three kinds of Ramanujan graphs. The degrees of these graphs are not bounded, yet they are very interesting graphs.

The first family is the family of Paley graphs:

Definition 8.3.1. *Let K be a finite field of order q, with $q \equiv 1 (\mathrm{mod}\, 4)$. Define a graph $X(q)$ whose vertices are the elements of K and where x is joined to y if $x - y$ is a nonzero square in K.*

It was observed by P. de la Harpe that the Paley graphs are Ramanujan. Indeed, these graphs are known to be strongly regular.

Definition 8.3.2. *A k-regular graph X with n vertices is called a strongly regular graph with parameters (n, k, r_1, r_2) if any two adjacent vertices in X have exactly r_1 common neighbors and any two non-adjacent vertices in X have exactly r_2 common neighbors.*

It is well known and easy to see directly that the adjacency matrix δ of a strongly regular graph X with parameters (n, k, r_1, r_2) satisfies $\delta^2 = (k - r_2)I + (r_1 - r_2)\delta + r_2 J$ where J is the $n \times n$ matrix whose all entries are equal to 1. The eigenvalues of X are therefore k (with multiplicity one if $r_2 > 0$) and the two roots of $f(x) = x^2 - (r_1 - r_2)x - (k - r_2)$. Thus X is a Ramanujan graph if and only if $3k - 4 + r_2 \geq 2|r_1 - r_2|\sqrt{k-1}$. Now, it is known that the Paley graphs $X(q)$ are strongly regular (cf. [HP, §3.7]) with parameters $(q, \frac{q-1}{2}, \frac{q-5}{4}, \frac{q-1}{4})$, and therefore we can deduce:

Proposition 8.3.3. *The Paley graphs $X(q)$ are Ramanujan graphs.*

Another family of graphs with such a number theoretic flavor was defined by A. Terras:

Definition 8.3.4. *Let $\mathbb{F}_q$ be a finite field of order q and let $\delta \in \mathbb{F}_q$ a nonsquare. Fix an element $a \in \mathbb{F}_p$ with $a \neq 0, 4\delta$. Define a graph $H_q(\delta, a)$ in the following way: its vertices are the points of the finite upper half-plane $\{z = x + y\sqrt{\delta} | x, y \in \mathbb{F}_q, y \neq 0\}$. Two points $z, \omega \in H_q$ are connected by an edge if $d(z, \omega) = a$ where $d(z, \omega) = \frac{N(z-\omega)}{Im(z)Im(\omega)}$, $N(x + y\sqrt{\delta}) = x^2 + \delta y^2$ and $Im(x + y\sqrt{\delta}) = y$.*

In a series of papers (cf. [Te2], [CPTTV] and [ACPTTV] and the references therein), A. Terras and her students studied the graphs $H_q(\delta, a)$, which are usually called by abuse of the language « the finite upper half plane». They are $(q + 1)$-regular graphs. Terras conjectured that they are Ramanujan, and this was proved recently to be the case:

Theorem 8.3.5. (N. Katz [Kat2]) *The graphs $H_q(\delta, a)$ are $(q + 1)$-regular Ramanujan graphs.*

The third family of Ramanujan graphs we mention here are the «generalized N-gons». These objects are usually treated as incidence planes of points and line but can actually be thought as bi-partite graphs. The conditions on such an object to be a «generalized N-gons» are strong enough to give their eigenvalues even though their structure and classification are not known. The main result in this topic is a theorem of Feit and Higman which asserts that there are only finitely many N's for which a non-trivial generalized N-gon exists (i.e., $N = 2, 3, 4, 6, 8, 12$). We will see that some of these generalized N-gons give Ramanujan graphs, but the Feit-Higman Theorem eliminates any hope of getting infinitely many Ramanujan k-regular graphs (for a fixed k!) using this method. Still these graphs have very strong bounds on their eigenvalues and might be useful.

We begin with presenting them by using mainly the language of graphs. We follow Feit-Higman [FH] and Tanner [Ta].

The generalized polygons are incidence structures consisting of points and lines. We will restrict our attention to those in which every point is incident to $r + 1$ lines and every line is incident to $r + 1$ points for some positive integer r. By identifying points with input vertices and lines with output vertices, a generalized N-gon defines a bi-partite graph X.

Definition 8.3.6. *A generalized N-gon is a bi-partite $k = r + 1$ regular graph X satisfying:*

(i) *For all vertices $x, y \in X$, $\mathrm{dist}(x, y) \leq N$ (i.e., $\mathrm{diam}(X) \leq N$).*

(ii) *If $\mathrm{dist}(x, y) = h < N$, then there exists a unique path of length h joining x and y (i.e., girth$(X) \geq 2N$).*

(iii) *Given $x \in X$, there exists a vertex $y \in X$ such that $\mathrm{dist}(x, y) = N$.*

Example 8.3.7.

(A) If $r = 1$ then X is just the standard $2N$-gon.

(B) If $N = 2$, then X is the complete bi-partite k-regular graph (i.e., every point being incident with every line).

(C) If $N \geq 3$, there is at most one line incident with any two distinct points. For $N = 3$, there is exactly one line incident with any two distinct points, and X is a projective plane (or more precisely, the points versus lines graph of a projective plane).

It was shown by Feit and Higman [FH] that the conditions of (8.3.6) suffice to determine the characteristic polynomial of $M = A^t A$ when A is the incidence matrix of X, i.e., the matrix whose rows are indexed by the lines (outputs) of X and whose columns are indexed by the points (inputs) and which has 1 in the place (e, f) if e is incident with f and 0 otherwise. In our standard notations the adjacency matrix δ of X is equal to $\begin{pmatrix} 0 & A^t \\ A & 0 \end{pmatrix}$. Hence $\delta^2 = \begin{pmatrix} A^t A & 0 \\ 0 & AA^t \end{pmatrix}$. The method of Feit-Higman enables one to compute the eigenvalues of $A^t A$ and hence of δ^2. Thus, up to sign one can also compute the eigenvalues of δ. In [Ta], Tanner gives the list of eigenvalues of generalized N-gons for $N = 3, 4, 6, 8$. (By the Feit-Higman Theorem these are the only interesting cases.) The list is as follows:

N	Eigenvalues of $A^t A$
3	$(s+1)^2, s$
4	$(s+1)(r+1),\ s+r, 0$
6	$(s+1)(r+1),\ r+s+\sqrt{rs},\ r+s-\sqrt{rs},\ 0$
8	$(s+1)(r+1),\ r+s+\sqrt{2rs},\ r+s,\ r+s-\sqrt{2rs},\ 0.$

In the above table we used the standard notation which allows the degree of the inputs to be $s + 1$ and the outputs $r + 1$. In the regular case, which is our main interest, we simply take $r = s$. (In this case $(r = s)$, $N = 8$ cannot actually occur.)

These are the eigenvalues for δ^2. From it one can easily calculate (up to sign) the eigenvalues of δ. This is an easy calculation which is left to the reader. The outcome is a nice surprise!

Theorem 8.3.8. *Regular (i.e., $r = s$) generalized N-gons are Ramanujan bi-partite graphs.*

As mentioned above, the Feit-Higman Theorem puts a limit on the usefulness of this result. Still there are quite a few generalized N-gons (see [Ta], [FH] and the references therein).

The discussion above calls attention to the following problem:

Problem 8.3.9. *Call a bi-partite graph X, a (r, s)-regular graph if the degree of all the inputs is $r + 1$ and the degree of all the outputs is $s + 1$. Define a Ramanujan (r, s)-regular graph (i.e., find the optimal bound on its eigenvalues).*

Are generalized N-gons always Ramanujan? Find explicit constructions of (r,s)-regular Ramanujan graphs.

(The first two problems should not be too difficult (see [Ha] and [Gg]). The last one is probably a very difficult one. One may try to use the arithmetic quotient of the Bruhat-Tits tree associated with some rank-one semi-simple groups over local non-archimedian fields; cf. [BT]).

Another suggestive remark was made by Bill Kantor: The generalized N-gons as bi-partite graphs are exactly rank 2 buildings with Weyl group D_{2N}. Once again this calls attention to the possibility of the existence of «Ramanujan Buildings» or «Ramanujan complexes». At this point, it is not clear what should be the right definition of such an object.

8.4 Ramanujan Diagrams

The Ramanujan graphs constructed in Chapter 7 are quotients of type $\Gamma\backslash \mathrm{PGL}_2(F)/K$ where F is a local non-archimedian field, K a maximal compact subgroup and Γ a cocompact lattice in $\mathrm{PGL}_2(F)$. In case of $\mathrm{char}(F) = p > 0$, $\mathrm{PGL}_2(F)$ also has non-uniform lattices, i.e., discrete subgroups Γ for which $\Gamma \backslash \mathrm{PGL}_2(F)$ has a finite invariant measure but is not compact. In this case the quotient $\Gamma \backslash \mathrm{PGL}_2(F)/K$ has in a natural way a structure of a «diagram», i.e., an infinite graph with weights for the vertices and edges such that the total volume/weight is finite. In [Mor 2] and [Mor 3], M. Morgenstern developed the theory of such diagrams: expanders, their Laplacian and eigenvalues, Ramanujan diagrams and their construction, etc. Beside the intrinsic interest in such object, there has been an unexpected reward: some finite subgraphs of the Ramanujan diagrams constructed, turn out to be excellent «bounded concentrators» (see Definition (1.1.7)). Thus they can apply to improve some network constructions.

This section is devoted to a brief description of these exciting developments. Details and proofs can be found in [Mor 2] and [Mor 3].

Definition 8.4.1. *A* <u>diagram</u> *is a triple $D = \langle V, E, \omega\rangle$ where $Y = \langle V, E\rangle$ is an undirected countable graph, $\omega : V \cup E \to \{\frac{1}{n} | n = 1,2,3,\ldots\}$ is a weight function and for any $e = (u,v) \in E$, $\frac{1}{\omega(e)}$ divides $\frac{1}{\omega(u)}$ and $\frac{1}{\omega(v)}$.*

For $A \subseteq V$, denote $\mu(A) = \sum_{u\in A} \omega(u)$. μ is the measure on D, and we assume that $\mu(V) < \infty$. Call $\theta(u,v) = \omega(e)/\omega(u)$ the entering degree of $e = (u,v)$ to u and call D k-regular if for every $u \in V$, $\sum_{(u,v)\in E} \theta(u,v) = k$.

Example 8.4.2. Let X_k be the k-regular tree, and $G = \mathrm{Aut}(X_k)$. If $k = p^r + 1 = q + 1$, p a prime, then $H = \mathrm{PGL}_2(\mathbb{F}_q((\frac{1}{t})))$ is a subgroup of G, where $\mathbb{F}_q((\frac{1}{t}))$ is the Laurent power series in $\frac{1}{t}$ over $\mathbb{F}_q$. Let Γ be a non-uniform lattice in G or in H. Note that H is cocompact in G and hence a lattice in H is automatically a lattice in G (but, of course, not vice versa). Let $\mathfrak{g} = (\Gamma, \Gamma \backslash X_k)$ be the quotient graph of groups (see [S1, §I.5]), i.e., the quotient graph $\Gamma \setminus X_k$, where for every vertex $\bar{x} = \Gamma \cdot x$ of the quotient graph we associate the stabilizer $\Gamma_x = \{\gamma \in \Gamma \mid \gamma(x) = x\}$ and, similarly, Γ_e is associated to the edge $\bar{e} = \Gamma \cdot e$ of $\Gamma \setminus X_k$. We now define a structure of a diagram on $\Gamma \setminus X_k$ by: $\omega(\bar{x}) = \frac{1}{|\Gamma_x|}$ and $\omega(\bar{e}) = \frac{1}{|\Gamma_e|}$ where notations are as above. As Γ is discrete, Γ_x and Γ_e are finite, and it is not difficult to see that ω satisfies the required condition. Moreover, $\mu(\Gamma \setminus X_k) = \tilde{\mu}(\Gamma \setminus G)$ where $\tilde{\mu}$ is the Haar measure of G normalized so that $\tilde{\mu}(K) = 1$ where K is a maximal compact subgroup of G, i.e., a stabilizer of one vertex. See [S1, p. 84] for more details. Few detailed examples of $\Gamma \setminus X_{q+1}$ for Γ-arithmetic subgroups of $\mathrm{PSL}_2(\mathbb{F}_q((\frac{1}{t})))$ are presented in [S1, II.2]. For $\Gamma = \mathrm{PSL}_2(\mathbb{F}_q[t])$ one gets a particularly simple picture: $T = \Gamma \setminus X_{q+1}$ is one infinite half ray.

$$T = \overset{\Lambda_0}{\bullet}\!-\!\!-\!\!-\overset{\Lambda_1}{\bullet}\!-\!\!-\!\!-\overset{\Lambda_2}{\bullet}\!-\!\!-\!\!-\overset{\Lambda_3}{\bullet}\!-\!\!-\!\!-\bullet\!-\!\!-\!\!- \quad \cdots\cdots$$

This ray can be lifted to the tree X_{q+1} in such a way that Λ_n is represented by the lattice spanned by $\{t^n e_1, e_2\}$ where $\{e_1, e_2\}$ is the standard basis of $\mathbb{F}((\frac{1}{t}))^2$. (See the description in 5.3 of the tree associated with $\mathrm{PGL}_2(\mathbb{Q}_p)$; a similar description applies to $\mathrm{PGL}_2(\mathbb{F}_q((\frac{1}{t})))$; see [S1]). The stabilizer of Λ_n in Γ is therefore Γ_n where:

$$\Gamma_0 = \mathrm{PSL}_2(\mathbb{F}_q) \text{ and for } n \geq 1,$$

$$\Gamma_n = \{\begin{pmatrix} a & b \\ 0 & d \end{pmatrix} \mid a, d \in \mathbb{F}_q \setminus \{0\},\ b \in \mathbb{F}_q[t] \text{ and } \deg(b) \leq n\}.$$

We therefore have $\omega(\Lambda_0) = |\mathrm{PSL}_2(\mathbb{F}_q)|^{-1}$ and $\omega(\Lambda_n) = ((q-1)q^{n+1}/\varepsilon_q)^{-1}$ when $\varepsilon_q = 2$ if q is odd and $\varepsilon_q = 1$ if q is even.

This diagram is especially simple. The other examples described in [S1] are not that simple, but all have a common feature: they are a union of a finite graph plus finitely many infinite rays (cusps). In [Lu2] it is shown that all non-uniform lattices of $H = \mathrm{PGL}_2(\mathbb{F}_q((\frac{1}{t})))$ have that property. On the other hand, examples given in [BL] show that this is not the case for arbitrary lattices in $G = \mathrm{Aut}(X_k)$. The corresponding diagrams can have an infinite homology, infinitely many cusps, etc. So general k-regular diagrams are quite wild!

If D is a diagram then the inner product of functions on D is defined by:

$$\langle f, g \rangle = \int_D f \cdot \bar{g} d\mu = \sum_{v \in V} f(\nu)\overline{g(\nu)}\omega(\nu)$$

and $\|f\|$ and $L^2(D)$ are as usual. The adjacency operator δ is defined by $\delta(f)(u) = \sum_{(u,v) \in E} \theta(u, v) \cdot f(v)$. In this section we call δ the Laplacian of D.

For simplicity of exposition, assume that D is k-regular and bi-partite, $V = I \bigcup O$ and assume $\mu(I) = \mu(O)$. Let $L_0^2(D) = \{f \in L_2(D) \mid \sum_{v\in I} f(v)\omega(v) = \sum_{v\in O} f(v)\omega(v) = 0\}$ and $\lambda(D) = \sup \operatorname{Spec}(\delta\mid_{L_2^0(D)})$, i.e., the norm of the operator δ restricted to $L_0^2(D)$.

One can define expanders in the context of diagrams in a natural way by taking the weight of a set and its boundary into consideration. Morgenstern [Mor2] did so and also developed the connection between this and $\lambda(D)$ in analogy with Propositions 4.2.4 and 4.2.5. (It should be mentioned, however, that this analogy is not straightforward and in fact some open questions were left. We will not go into details here, referring the reader to [Mor2] for more.) He also showed the following analogue of (4.5.5):

Proposition 8.4.3. *For a k-regular infinite diagram D, $\lambda(D) \geq 2\sqrt{k-1}$.*

Note that here we do not need an infinite family and that the result holds for any single *infinite* diagram. The following definition is therefore natural:

Definition 8.4.4. *An infinite k-regular diagram with $\lambda(D) = 2\sqrt{k-1}$ is called a Ramanujan diagram.*

Example 8.4.5. Let $\Gamma = \mathrm{SL}_2(\mathbb{F}_p[t])$ and D the quotient diagram of $\Gamma \setminus X_{p+1}$ described in Example 8.4.2. Then D is a Ramanujan diagram. For a detailed explicit computation of the spectrum of δ on D; see Efrat [Ef].

In [Dr2] Drinfeld proved the function field analogue of Ramanujan conjecture and Selberg's Theorem (compare Corollaries 5.5.2 and 5.5.3). From this Morgenstern deduced:

Theorem 8.4.6. *Let Γ' be a congruence subgroup of $\Gamma = \mathrm{PSL}_2(\mathbb{F}_q[t])$. The diagram $\Gamma' \setminus X_{q+1} = \Gamma' \setminus \mathrm{PGL}_2(\mathbb{F}_q((\frac{1}{t})))/\mathrm{PGL}_2(\mathbb{F}_q[[\frac{1}{t}]])$ is a Ramanujan diagram.*

For example, if $g = g(t) \in \mathbb{F}_q[t]$ is a polynomial of degree $d \geq 2$ and $R = \mathbb{F}_q[t]/(g(x))$, $\Gamma' = \Gamma(g(t)) = \mathrm{Ker}(\mathrm{PSL}_2(\mathbb{F}_q[t]) \to \mathrm{PSL}_2(R))$ then $D_g = \Gamma' \setminus X_{q+1}$ is a finite sheeted (ramified) cover of $D = \Gamma \setminus X_{q+1}$, where $\Gamma = \mathrm{PSL}_2(\mathbb{F}_q[t])$. The latter is described in Example 8.4.2. For $i = 0, 1, 2, \ldots$, let L_i be the set of vertices of D_g lying above Λ_i – call them the vertices of *level i*. Then the vertices in L_i are connected only with vertices from L_{i-1} and L_{i+1}. The group Γ/Γ' acts transitively on L_i and the set L_i can be, therefore, identified with $\Gamma/\Gamma_i\Gamma'$, where Γ_i is the stabilizer of Λ_i in Γ (see Example 8.4.2). This actually can be translated to a very explicit construction: Let R be represented by the polynomials of degree smaller than d, $M = \mathrm{PSL}_2(R)$, $M_0 = \mathrm{PSL}_2(\mathbb{F}_q)$ and for $i > 0$:

$$M_i = \{\begin{pmatrix} a & b \\ 0 & d \end{pmatrix} \in M \mid a, d \in \mathbb{F}_q \setminus \{0\}, \deg(b) \leq i\}.$$

D_g has the following description: The vertices L_i of level i are the elements of M/M_i for $i = 0, 1, 2 \ldots$, and an edge exists between xM_i and yM_{i+1} if and only if their intersection is *not* empty. From this description we see that for $i \geq d-1$, $|L_i| = |L_{d-1}|$ and we have $|L_d|$ cusps in D_g.

Diagram D_g

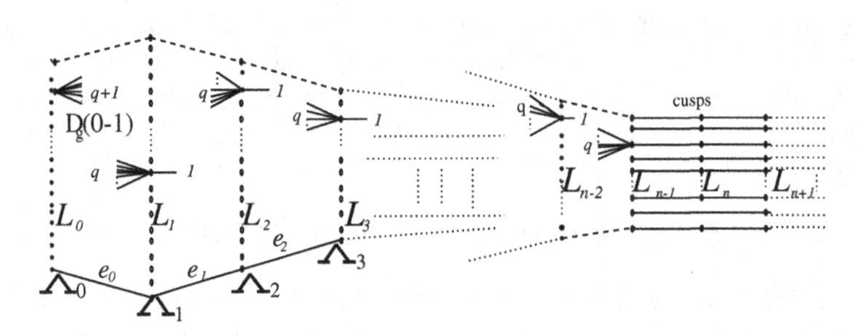

Now if we restrict our attention only to levels 0 and 1, we get a finite bi-partite graph (the weights here are all 1), with $|L_0| = \frac{q^{3d}}{q(q^2-1)}\Psi$ and $|L_1| = \frac{q^{3d}}{q^2(q-1)}\Psi$ where Ψ is equal to $\prod_{i=1}^{s}(1 - \frac{1}{q^{2r_i}})$ where r_i are as follows: write $g(t) = p_1(t)^{\alpha_1} \cdots p_s(t)^{\alpha_s}$ where $p_i(t)$ are irreducible of degree r_i (so $\sum_{i=1}^{s} \alpha_i r_i = d$). Anyway L_1 is somewhat larger than L_0. Morgenstern noticed that applying the fact that D_g is Ramanujan to the expansion properties of the subgraph $L_0 \cup L_1$ yields:

Proposition 8.4.7. *If $q \geq 5$ then the subgraph $L_0 \cup L_1$ of D_g is a $\left(\frac{q^{3d}}{q^2(q-1)}\Psi, \frac{q}{q+1}, q, \frac{q-4}{q-3}\right)$-bounded concentrator (see Definition (1.1.7)).*

These bounded concentrators have some advantages over the one described in Chapter 1, e.g., homogenicity. They give rise to superconcentrators of density 66. Proposition 8.4.7 does not hold for $q = 2$, but if one can prove it for $q = 3$ it would give superconcentrators of density 28. This would be better than what can be done by random methods!

9 Distributing Points on the Sphere

9.0 Introduction

A problem of practical importance is that of generating a large number of random elements in a group. Of particular interest are the orthogonal groups $SO(n+1)$. A closely related problem is the one of placing a large number of points on the sphere S^n in a uniform way. For a survey and many references see Sloane [Sl] and Berger [Beg].

In this chapter, we present a solution for this problem for the case $n = 2, 3$. This is a method based on previous considerations and in particular on Deligne's Theorem (6.1.2). It has some remarkable properties which are impossible to be achieved for $n = 1$. Just like with the Banach-Ruziewitz problem, this is because of the amenability of S^1. For $n \geq 4$, on the other hand, only weaker properties are known to exist. The problem is still widely open.

The results presented in this chapter give a method to distribute points on the sphere which is satisfactory from one point of view. There are many different criterions and methods depending on need. For information on other methods see the extensive bibliography of [CS]. The results here are due to Lubotzky-Phillips-Sarnak [LPS3, LPS4]. The presentation, however, partially follows the exposition of those results by Colin de Verdiere [CV3].

9.1 Hecke operators of group action

In this section we introduce «Hecke operator» δ_S in the context of harmonic analysis of a compact space X, where S is a finite subset of isometries of X. We relate the eigenvalues of δ_S to the distribution of the points Sx_0 for $x_0 \in X$.

So let $S = \{a_1, b_1 = a_1^{-1}, \ldots, a_\ell, b_\ell = a_\ell^{-1}\}$ be a finite symmetric subset of isometries of X, and let for $f \in L^2(X)$, $(\delta_S f)(x) = \sum_{\gamma \in S} f(\gamma x)$. It is not difficult to check that the operator δ_S is symmetric, of norm 2ℓ and $\delta_S(1) = 2\ell \cdot 1$ where 1 is the constant function. The subspace $L_0^2(X) = \{f \mid \int f = 0\}$ is invariant under δ_S. Let

$$\lambda(S) = \frac{1}{2\ell} \|\delta_S\ |_{L_0^2(X)}\ \|.$$

So, $2\ell\ \lambda(S)$ is the norm of δ_S restricted to $L_0^2(X)$.

The ultimate goal is to find a set S for which $(Df)(x) = \frac{1}{2\ell} \sum_{\gamma \in S} f(\gamma x) - \frac{1}{\mathrm{vol}(X)} \int f d\mu$ is as small as possible in an appropriate norm. Once we achieve this goal, we can use the finite sum $\frac{\mathrm{vol}(X)}{2\ell} \sum_{\gamma \in S} f(\gamma x)$ as a good approximation for $\int f d\mu$. Now, $Pf = \frac{1}{\mathrm{vol}(X)} \int f d\mu$ is the projection of f to the space of constants, so

$f - Pf \in L_0^2(X)$ and $Df = \frac{1}{2\ell}\delta_S(f - Pf)$. So $\lambda(S) = \|D\|$ when we consider D as an operator for $L^2(X)$ to $L^2(X)$. Our goal is, therefore, to find S for which $\lambda(S)$ is as small as possible. (So we are looking for optimal bounds in the L^2-norm, but eventually we also have estimation with the L^∞ norm, i.e., for every point, and not just on the average in the L^2-norm, see (9.2.6) below.)

Lemma 9.1.1. *In the notations above, $\lambda(S) < 1$ if and only if the representation of the finitely generated group $\Gamma = \langle S \rangle$, generated by S, on $L_0^2(X)$ does not weakly contain the trivial representation.*

We omit the proof of this lemma, which is analogous to the proof of Propositions 4.2.4 and 4.2.5 (compare also (4.3.2)). One can, in fact, think of X here as a graph in which every $x \in X$ is connected to $\gamma(X)$ for every $\gamma \in S$. The graph is certainly not connected, so on the face of it (recalling that for a finite graph the trivial eigenvalue has multiplicity one if and only if the graph is connected) one does not expect $\lambda(S)$ to be smaller than one. Still, there is a difference: The connected component of $x \in X$ is the orbit Γx when $\Gamma = \langle S \rangle$. So if one orbit is dense, 1 is not an eigenvalue of δ_S on $L_0^2(X)$. It is still a difficult task to make sure that the norm of δ_S on $L_0^2(X)$ is strictly less than 1.

The lemma gives:

Proposition 9.1.2. *Let Γ be a discrete amenable group acting on a compact space X and $S \subseteq \Gamma$ a symmetric subset with 2ℓ elements. Then $\lambda(S) = 1$.*

Proof. The group generated by S is also amenable, and whenever an amenable group acts on X, its representation on $L_0^2(X)$ weakly contains the trivial representation (compare (3.1.5)). □

On the other hand:

Proposition 9.1.3. *If $\Gamma = \langle S \rangle$ is a Kazhdan group acting ergodicly on X, then $\lambda(S) < 1$.*

Proof. Because of the ergodicity, there is no Γ-invariant function in $L_0^2(X)$. By the definition of Kazhdan group (§3.1), its representation on $L_0^2(X)$ does not weakly contain the trivial representation. □

Proposition 9.1.2 implies that for any finite subset S of $O(2)$ acting on the circle S^1, $\lambda(S) = 1$. On the other hand, just like the Ruziewicz problem and because of precisely the same reasoning the situation for S^n, $n \geq 2$ is different:

Theorem 9.1.4. *For every $n \geq 2$ there exists $\ell = \ell(n)$ and $\varepsilon = \varepsilon(n) > 0$ such that $SO(n+1)$ has a symmetric finite subset S of order 2ℓ for which $\lambda(S) \leq 1 - \varepsilon$, where $\lambda(S)$ is the norm of δ_S on $L_0^2(S^n)$.*

Proof. In Lemma 3.4.3 we proved that $SO(n)$, $n \geq 5$ contains a finitely generated subgroup Γ which has property (T). Taking S to be a symmetric set of generators for Γ, proves the result for $n \geq 4$.

For $n = 2, 3$ we have produced in the proof of Theorem 7.2.1, a finitely generated group (in fact a free group, at least for $n = 2$) of $SO(n+1)$, which does not have property (T), but its action on $L_0^2(S^n)$ does not weakly contain the trivial representation. A set S of generators for Γ will do it here. □

In the next section we shall present a set S of 6 elements for which $\lambda(S) = 2\sqrt{5}$. Moreover, Sarnak [Sa2] shows how, given a set of $2\ell(n)$ elements in $SO(n+1)$ with $\lambda(S) < 1 - \epsilon(n)$, one can find in an effective way a set S' of $4\ell(n) = 2\ell(n+1)$ elements in $SO(n+2)$ with $\lambda(S') < 1 - \epsilon(n+1)$. An open problem is to find such sets with ℓ and ε independent of n. It is quite likely that ℓ can be made independent of n (by finding cocompact lattices in $SO(n-2, 2)$, for $n \to \infty$, as in the proof of Proposition 3.4.3, with a bounded number of generators). On the other hand, it is an open problem whether using bounded ℓ, ε can be made independent of n. This interesting question might be related to the problem, posed in Example 4.3.3C of making all the symmetric groups S_m a family of k-regular expanders, simultaneously.

9.2 Distributing points on S^2 (and S^3)

We first present a lower bound of how small $\lambda(S)$ can be. We then, following [LPS1, LPS2], present a set of elements which achieves this optimal bound.

Theorem 9.2.1. *For every $\ell \in \mathbb{N}$ and every symmetric subset of $SO(3)$ of order 2ℓ acting on $X = S^2$, $\lambda(S) \geq \frac{2\sqrt{2\ell-1}}{2\ell} = \frac{\sqrt{2\ell-1}}{\ell}$.*

Proof. See [LPS3, Theorem 1.3] or [CV3, Theorem A].

This theorem is an analogue of the bound (4.5.4) and (4.5.5) for graphs. Its proof actually also uses (4.5.3) with some information on spherical harmonics on S^2. It seems likely that there should be a general result like (9.2.1) for much more general X.

Our favorite group Γ, which was presented first under the name $\Lambda(2)$ in (2.1.11) and have been used also in §7.4 to achieve the optimal bound for graphs (Ramanujan graphs) is the one which will give the optimal $\lambda(S)$ here. But first a definition:

Definition 9.2.2. *A symmetric subset S of size 2ℓ of $SO(3)$ is called a Ramanujan set if with respect to its action on S^2, $\lambda(S) \leq \frac{\sqrt{2\ell-1}}{\ell}$.*

Theorem 9.2.3. *Let p be a prime, $p \equiv 1 (\mathrm{mod}\, 4)$. Let*

$$S' = \{\alpha = (a_0, a_1, a_2, a_3) \in \mathbb{Z}^4 \mid \sum_{i=0}^{3} a_i^2 = p, a_0 > 0 \text{ odd},$$

$$\text{and } a_i \text{ even for } i = 1, 2, 3\}.$$

To every $\alpha \in S'$ associate an $\alpha' \in PGL_2(\mathbb{C})$,

$$\alpha' = \begin{pmatrix} a_0 + a_1 i & a_2 + a_3 i \\ -a_2 + a_3 i & a_0 - a_1 i \end{pmatrix}.$$

α' is in fact in $PU(2)$, which is isomorphic to $SO(3)$. The isomorphism φ may be obtained by identifying $\mathbb{R}^3$ with real quaternions of the form $\{x_i + y_i + zk \mid (x, y, z) \in \mathbb{R}^3\}$ and letting α acting on it by conjugation. Let $\tilde{\alpha} = \varphi(\alpha')$; then the set $S = \{\tilde{\alpha} \mid \alpha \in S'\}$ is a symmetric Ramanujan set of order $p + 1 = 2\ell$, i.e., $\lambda(S) \leq \frac{2\sqrt{p}}{p+1}$.

Proof. That $|S| = p + 1$ follows from 2.1.8 (see also (2.1.11) and §7.4). To prove the Ramanujan bound, one argues in a way similar to the proof of (7.2.1). The current theorem can actually be considered as a quantitative version of (7.2.1). □

Example 9.2.4. Let $p = 5$ and so $S' = \{\alpha_1, \alpha_2, \alpha_3, \bar{\alpha}_1, \bar{\alpha}_2, \bar{\alpha}_3\}$ where $\alpha_1 = (1, 2, 0, 0)$, $\alpha_2 = (1, 0, 2, 0)$ and $\alpha_3 = (1, 0, 0, 2)$ (and $\bar{\alpha}$ is in the quaternionic sense, i.e., if $\alpha = (x, y, z, w)$ then $\bar{\alpha} = (x, -y, -z, -w)$). Then $S = \{\tilde{\alpha}_1, \tilde{\alpha}_2, \tilde{\alpha}_3, \tilde{\alpha}_1^{-1}, \tilde{\alpha}_2^{-1}, \tilde{\alpha}_3^{-1}\}$ where $\tilde{\alpha}_1$, $\tilde{\alpha}_2$ and $\tilde{\alpha}_3$ are rotations of $\arccos(-\frac{3}{5})$ about the X, Y, Z-axes, respectively (since $(1 + 2i)j(1 - 2i) = 3j + 4k$, etc.). These are rotations of angle approximately $126°52'$ around the axes.

It should be mentioned that most subsets of $SO(3)$ are not Ramanujan. The following theorem is proved in [LPS1]:

Theorem 9.2.5. *For generic $\underline{g} = (g_1, \ldots, g_\ell) \in SO(3)^\ell$, the set $S = \{g_1^{\pm 1}, \ldots, g_\ell^{\pm 1}\}$ is not a Ramanujan set. (By «generic» we mean that the compliment, i.e., the ℓ-tuples which are Ramanujan, form a subset of the first category.)*

It is interesting to compare this with J. Friedman's results ([Fr]), which show that most k-regular graphs are almost Ramanujan (and it is conjectured that they are indeed Ramanujan). They are, anyway, expanders by Proposition 1.2.1, while in the current situation it is an open problem whether for generic S, $\lambda(S) < 1$ or $\lambda(S) = 1$.

Finally we bring some consequences which are of importance for numerical analysis:

For a subset S of $SO(3)$ and $N \in \mathbb{N}$, let ${}^N S = \{\gamma \in SO(3) \mid \gamma \text{ is a word of length} \leq N \text{ in } S\}$.

Theorem 9.2.6. *If S is a Ramanujan subset of $SO(3)$, then $\lambda({}^N S) = O(\log n/\sqrt{n})$ when $n = |{}^N S|$.*

Proof. See [LPS3, Theorem 1.5] or [CV3, Theorem C]. □

The theorem implies that for every $f \in L^2(S^2)$, $\|\frac{1}{n}\sum_{j=1}^{n} f(\gamma_j x) - \frac{1}{4\pi}\int_{S^2} f\|_{L^2} = O(\log n/\sqrt{n})$ when $\{\gamma_j \mid 1 \leq j \leq n\}$ is the set of elements in ${}^N S$.

When f is of somewhat special form one can also get bounds in L^∞, i.e., for every $x \in S^2$.

Theorem 9.2.7. *Suppose S is a Ramanujan subset of $SO(3)$, D a domain in S^2 with a piecewise smooth boundary and x_0 any point of S^2, then:*

$$\left|\frac{1}{n}\#\{j \mid \gamma_j x_0 \in D\} - \frac{1}{4\pi}\text{Area}(D)\right| = 0(\log n/n^{1/3}),$$

where $\{\gamma_j \mid 1 \leq j \leq n\}$ is the set of elements in ${}^N S$.

Proof. [LPS3, Theorem 2.7] or [CV3, Theorem D]. □

Such a theorem is of practical importance: it gives an algorithm to compute numerically an area of a subset D of S^2. This will be done by picking any point x_0 of S^2 and applying on it the words of length $\leq N$ in the $A^{\pm 1}, B^{\pm 1}$ and $C^{\pm 1}$ of Example 9.2.4. The proportion of them which lie in D will be an estimation to the area of D where we also have an estimate of the rate of convergence of the algorithm. One of the significance properties of this algorithm is that one does not have to decide in advance on N (and n) but rather can stop the algorithm while it achieves the degree of precision needed. The design of the algorithm is not at all dependent on the number of points which are going to be sampled!

Remark 9.2.8. It is indeed proved in Theorem 9.2.3 that the set S constructed there satisfies the Ramanujan bound also with respect to its action on $L_0^2(SO(3))$ and even on $L_0^2(SU(2))$ when S is considered as a subset of $SU(2)$ instead of $SO(3)$ (by replacing $\tilde{\alpha}$ by α'). As $SU(2)$ is topologically the three-dimensional sphere, the results in this section gives also an effective method to distribute points on S^3.

10 Open Problems

The results presented in this book solve some problems and at the same time open many others. In this chapter we will collect some of these and try along the way to give some indications of what we believe the questions which future research should face. We categorize the problems in accordance to the chapters of the book, but this is somewhat arbitrary as many of them are related to more than one chapter.

10.1 Expanding graphs

The results presented in this book do not give a complete satisfactory solution to the problem of expanders. The Ramanujan graphs are excellent expanders but not as good as what can be obtained by random methods. The best estimation of the expansion coefficient of Ramanujan graphs is due to Kahale [K1], [K2], who also showed that his estimation is the best possible if just a bound on λ_1 is given. One should therefore ask:

Problem 10.1.1. *What is the best expansion coefficient one can obtain for an infinite family of k-regular graph?*

Pippenger has some results in this direction. For the next question one needs a real new idea:

Problem 10.1.2. *Find expanders which will achieve this bound.*

It is also of interest:

Problem 10.1.3. *Find new methods to construct k-regular expanders for a fixed k.*

The problem of unbalanced expanders is also still widely open:

Problem 10.1.4. *Find an explicit construction of a bi-partite bi-regular graphs with vertex set $V = I \bigcup O$ such that all edges go between I and O, $|I| \sim n^2$, $|O| \sim n$, the degree of every vertex in I is bounded and such that for every $A \subseteq I$ of size $\leq \frac{n}{2}$, $|\partial A| \geq |A|$.*

There are various variants of this problem where random arguments ensure the existence of various types of graphs but explicit constructions are not known.

10.2 The Banach-Ruziewicz Problem

There are various interesting problems related with the Hausdorff-Banach-Tarski paradox. Those will not be mentioned here but rather refer the reader to Wagon [Wa]. We limit ourselves to one general problem and a special case of it:

Problem 10.2.1. *For what compact groups K, the Haar measure μ is the unique finitely additive K-invariant measure defined on the μ-measurable sets?*

It was shown by Margulis [M2] and Drinfeld [Dr1] (see Corollaries 3.4.6 and 7.2.2) that μ is the unique one when K is a non-abelian compact simple Lie group over $\mathbb{R}$. This is not the case for $SO(2)$, which is amenable as a discrete group. An interesting special case is:

Problem 10.2.2. *Let* $K = \prod_{n=5}^{\infty} A_n$ *where* A_n *is the alternating group on* n *elements. Answer Problem (10.2.1) for this* K.

The interest of this example is that $\prod_{n=5}^{\infty} A_n$ is not an amenable group as a discrete group; in fact, it contains the free group on 2 generators. (This can be deduced from the fact that the free group is residually-$\{A_n \mid n = 5, 6 \ldots\}$; see [Ma]). On the other hand, it is shown in [LW] that K contains a finitely generated dense amenable group. Compare also with Problems (10.3.4) and (10.3.6) below and see Sarnak [Sa2, p. 58].

10.3 Kazhdan Property (T) and its applications

Let Γ be a discrete group generated by a finite set of generators S, and let (H, ρ) be a unitary representation of Γ. Denote $K(\Gamma, S, \rho) = \inf_{\substack{v \in H \\ \|v\|=1}} \max_{s \in S} \{\|\rho(s)v - v\|\}$. So, $K(\Gamma, S, \rho) = 0$ if and only if ρ weakly contains the trivial representation. The group Γ has Kazhdan Property (T) if and only if there exists $\varepsilon(S) > 0$ such that $K(\Gamma, S, \rho) \geq \varepsilon(S)$ for every representation ρ of Γ which does not contain a nonzero Γ-fixed vector. The first question asks whether ε can be chosen independently of S.

Problem 10.3.1. *Let* Γ *be a discrete group with Kazhdan Property* (T). *Is there an* $\varepsilon > 0$ *such that* $K(\Gamma, S, \rho) > \varepsilon$ *for every set of generators* S *of* Γ *and every* ρ *which does not contain (strongly) the trivial representation?*

A positive answer will be quite surprising, but we can not eliminate this possibility. There are only few estimates of the Kazdhan constants $K(\Gamma, S, \rho)$ (see [Bur5] and [BDH]).

A problem of similar flavor:

Problem 10.3.2. *Let* $\{G_i\}$ *be a family of finite groups each with two systems of generators* S_i *and* S_i' *with* $|S_i|, |S_i'| \leq k$ *for every* i. *Assume the family of Cayley graphs* $X(G_i; S_i)$ *forms a family of expanders. Is the same true for* $X(G_i; S_i')$?

On the face of it the moral lesson of our discussions in Chapters 3 and 7 is that the question of being expanders highly depends on the chosen generators. Though in [LW] some evidence is presented that Problem 10.3.2 might have an affirmative answer. Such an answer would be truly remarkable.

A very interesting special case is:

Problem 10.3.3. *Is the family $\{SL_2(p) \mid p \text{ prime}\}$ a family of expanders with respect to all choices of generators? Some computations made in [LaRo] suggest that this might be the case or at least with respect to random generators. At this point we even do not know that the diameter of $SL_2(p)$ with respect to all (or at least random) sets of generators is $O(\log p)$. Compare also 10.8.*

The following can be also considered as a special case of Problem 10.3.2, but it is of great interest by itself:

Problem 10.3.4. *Can the family of symmetric groups S_n be made a family of expanders using a bounded number of generators?*

Of course if Problem 10.3.2 has a positive solution, then Problem 10.3.4 has a negative one since we know that $\{S_n\}$ is not a family of expanders with respect to $\tau = (1,2)$ and $\sigma = (1,2,\ldots,n)$. See Example 4.3.3C.

Recall that Alon and Roichman [AR] showed that every group is an expander with respect to $O(\log|G|)$ generators. In the case of S_n, we even do not know whether S_n gives a family of expanders using $O(n)$ generators.

A question of a similar flavor:

Problem 10.3.5. *For a fixed prime p, can the family $\{SL_n(p) \mid n = 2,3\ldots\}$ be made into a family of expanders using a bounded number of generators?*

Note that for a fixed n and variable p the family $\{\mathrm{SL}_n(p) | p \text{ prime}\}$ can be made into a family of expanders (by choosing generators of $\mathrm{SL}_n(\mathbb{Z})$; see Example 3.3.2 and Theorem 4.4.2i). As is shown in [LW], there is a difference between the family $\{\mathrm{SL}_n(p) \mid n \text{ fixed}, p \text{ varies}\}$ and $\{\mathrm{SL}_n(p) \mid p \text{ fixed}, n \text{ varies}\}$. For example, $\prod_n SL_n(p)$ contains a dense finitely generated amenable group, while $\prod_p SL_n(p)$ does not. The latter contains (for $n \geq 3$) a dense subgroup with Property (T). This is not known and probably not true for $\prod_n SL_n(p)$. It was conjectured in [LW] that the following problem has a positive answer:

Problem 10.3.6. *Let K be a compact group. Assume K contains two finitely generated dense subgroups A and B, such that A is amenable and B has Property (T). Is K necessarily finite?*

Compare also Problem 10.2.2 above and Problem 10.9.2 below. It seems that if one wants to try to give a negative answer to Problem 10.3.5 or a positive one to (10.2.2) and (10.9.2), then a new source of discrete subgroups with Property (T) is needed: Groups which are not lattices in semi-simple Lie groups. Finding such new examples seems to need some truly new idea (see [CMS]). It is also not known whether groups like the mapping class group of a surface of genus $g \geq 2$ or $\mathrm{Aut}(F_n)$ (or Out (F_n)) for $n \geq 3$ have Property (T) (where F_n is the free group on n generators). These groups have a lot in common with arithmetic groups; still, Property (T) for them is out of reach at this point.

10.4 The Laplacian and its eigenvalues

Selberg's Theorem (4.4.1) is very inspiring: recall that $\Gamma = \mathrm{SL}_2(\mathbb{Z})$ does not have the congruence subgroup property (i.e., it has a lot of finite index subgroups which do not contain $\Gamma(m) = \mathrm{Ker}(\mathrm{SL}_2(\mathbb{Z}) \to SL_2(\mathbb{Z}/m\mathbb{Z}))$ for any $m \neq 0$). At the same time $SL_2(\mathbb{Z})$ does not have Property (τ) with respect to the set of all finite index normal subgroups. Still if we ignore the non-congruence subgroups and «pretend» as only congruence subgroups exist, Property (τ) holds with respect to them. Moreover, for $\mathrm{SL}_2(\mathbb{Z}[\frac{1}{p}])$ the congruence subgroup property and (τ) both hold. If k is any number field and $\mathfrak{O}$ its ring of integers, then $\mathrm{SL}_2(\mathfrak{O})$ has the congruence subgroup property (i.e., $\mathrm{Ker}(\widehat{\mathrm{SL}_2(\mathfrak{O})} \to \mathrm{SL}_2(\widehat{\mathfrak{O}}))$ is finite) if and only if $\mathrm{SL}_2(\mathfrak{O})$ has Property (τ). It is quite natural therefore to conjecture that this is the situation in general:

Problem 10.4.1. *Let k be a global field, S a finite set of primes of k (including all the archimedean ones) and $\mathfrak{O}_S$ the ring of S-integers. Let G be a simply-connected, semi-simple algebraic group defined over k and $\Gamma = G(\mathfrak{O}_S)$. Prove that Γ has the congruence subgroup property (i.e. $\mathrm{Ker}(\widehat{G(\mathfrak{O}_S)} \to G(\widehat{\mathfrak{O}}_S)$ is finite) if and only if Γ has Property (τ).*

One side of this problem would follow once Problem 10.6.1 below would be proved. The more interesting direction is the other one. It would show in particular that Property (T) implies the congruence subgroup property (and hence also super-rigidity – see [BMS] and [Ra2]). It is interesting to compare this conjecture with Rapinchuk's conjecture which was proved recently ([PR], [Lu3]). This conjecture asserts that some purely group theoretic condition – $\Gamma = G(\mathcal{O}_S)$ is boundedly generated, i.e., a product of cyclic groups – implies the congruence subgroup property.

As a first step toward proving Problem 10.4.1, it will be useful to prove some group theoretic properties of groups with Property (τ), e.g.:

Problem 10.4.2. *Let Γ be a group with Property (τ). Prove some non-trivial bounds on various asymptotic invariants of Γ, e.g., the rate of growth of $d(H)$ with respect to the index of H in Γ (where H is a finite index subgroup of Γ and $d(H)$ is its number of generators). Or estimate $a_n(\Gamma)$ – the number of subgroups of Γ of index at most n (see [LW] for the first steps in this direction and compare [Lu3]).*

Problem 10.4.3. *Let Γ be a group with Property (τ). Prove that the pro-p completion $\Gamma_{\hat{p}}$ of Γ is an analytic group over some notherian pro-p ring. See [DDMS] and [LuS].*

In Section 4.5 we defined Ramanujan graphs, but only for the case of k-regular graphs. The following problem is quite natural:

Problem 10.4.4. *Extend the definition of Ramanujan graphs to arbitrary graphs and prove results analogue to Corollary 4.5.5 and the other results in §4.5. A Ramanujan graph should be a graph which «imitates» the universal covering tree of it. Does every locally finite tree which covers some finite graph also cover a Ramanujan graph? infinitely many Ramanujan graphs?*

Recall that in [BK] a necessary and sufficient condition is given for a tree to cover a finite graph. Note that at this point it is not even clear whether the k-regular tree, when $k-1$ is not a prime power, covers Ramanujan graphs. See Problem 10.7.3. The first steps in answering Problem 10.4.4 are given in [Gg]. See also [FTN], [FTP] and the references therein for harmonic analysis on trees. The case of immediate interest is (r, s)-bi-regular graphs. In this case there is even a reasonable chance to make an explicit constructions of Ramanujan graphs for some special values of r and s. This probably can be done by replacing $SL_2(\mathbb{Q}_p)$ in Chapter 7, by other rank one groups whose associated Bruhat-Tits buildings are bi-regular (and not regular) trees.

10.5 The representation theory of PGL_2

The material of Chapter 5 is a quite standard one. We have nothing to add in this direction.

10.6 Spectral decomposition of $L^2(G(\mathbb{Q}) \setminus G(\mathbb{A}))$

Chapter 6 also only brings some results from the literature. However, we call the attention of the reader to Conjecture 6.1.3. there. This is a very difficult task. The following seems to be a more realistic goal:

Problem 10.6.1. *Let K be a local non-discrete field, G a simply-connected, (almost) simple algebraic group defined over K and Γ an arithmetic lattice in $G(K)$. Assume K-rank$(G) = 1$. Prove that Γ has Property (τ) with respect to the family of its congruence subgroup (see Definition (4.3.1)).*

This is a version of the analogue of Selberg's Theorem for all rank one groups. For the case $K = \mathbb{R}$, this was already done by Elshodt, Gruenwald and Mennicke [EGM], Li, Piatetski-Shapiro and Sarnak [LPSS] and Li [Li]. Also for $G = \mathrm{SL}_2$ this is known over all fields due to the work of Selberg, Deligne, Drinfeld and Jacquet-Langlands. But the other examples of rank one groups over non-archimedian fields still need to be worked out. Doing this will give the following corollary: Let K, G and Γ be as above, without the assumption that G has rank one. Let S be a finite set of generators of Γ (such a set exists unless $\mathrm{char}(k) = p > 0$ and $\mathrm{rank}(G) = 1$, see [Lu2]). Then the family $\{X(\Gamma/N; S) \mid N$ a congruence normal subgroup of $\Gamma\}$ is a family of expanders.

It is worth notice that this corollary would essentially include all the known methods to construct expanders (see [LW]).

10.7 Banach-Ruziewicz problem for $n = 2, 3$; Ramanujan graphs

Experience with Ramanujan graphs shows that they « tend» to have high girth-but in all cases this was proved by direct methods and not as a corollary to the bound on the eigenvalues. A (very) partial explanation to this is Proposition 4.5.7, but it merely implies that the girth can not be bounded in some circumstances.

Problem 10.7.1. *Clarify the connection between eigenvalues and girth. Do Ramanujan graphs necessarily have logarithmic girth?*

A question of independent interest is:

Problem 10.7.2. *Fix k and let*

$$g(k) = \limsup \left\{ \frac{\mathrm{girth}(X_{n,k})}{\log_{k-1}(n)} \mid X_{n,k} \text{ is a } k- \text{ regular graph with } n \text{ vertices} \right\}.$$

It is easy to see that $g(k) \leq 2$, and Theorem 7.3.12 and the remark after Theorem 7.4.6 show that for $k = \ell+1$ where ℓ is a prime power, $g(k) \geq \frac{4}{3}$. Give better upper or lower estimates? Here one does not necessarily need explicit constructions.

Similarly, in the following question we do not ask for explicit constructions, though such one will be most desirable:

Problem 10.7.3. *Is it true that for every $k \geq 3$ there exist infinitely many k-regular Ramanujan graphs?*

Theorems 7.4.5 and 7.4.6 give a positive answer for every $k = p^r + 1$ where p is a prime and r any positive integer. But nothing is known for other k's. The first open problem is $k = 7$.

Problem 10.7.4. *Are random k-regular graphs Ramanujan? see [BS], [Fr1] and [FKS] where it is proved that they are «almost» Ramanujan.*

Of course it will be of great interest to find more methods to construct k-regular Ramanujan graphs for k fixed.

10.8 Some more discrete mathematics

When discussing diameter of a family of finite groups with respect to a bounded set of generators, there are three different issues to be considered: best generators, random generators and worst case generators.

For the family of finite simple non-abelian groups Theorem 8.1.4 says that «best generators» indeed generate them in $O(\log |G|)$ steps. We can still ask:

Problem 10.8.1. *Can the number seven in Theorem 8.1.4 be dropped to two?*

It was shown by Kantor [Kan1,2] that this is indeed the case for many of the simple groups.

Before passing to random generators, let us re-mention some problems on the generators we used in §8.1.

Problem 10.8.2. *Let $\{A_n, B_n\}$ be the standard set of generators of $\mathrm{SL}_n(\mathbb{Z})$ described in Proposition 8.1.1. Does there exist a constant C such that for every n and p, diam$X(\mathrm{SL}_n(p); \{A_n, B_n\}) \leq Cn^2 \log p$?*

It was observed [LW] that for a fixed p the family $X(\mathrm{SL}_n(p), \{A_n, B_n\})$ is not a family of expanders. It is a family of expanders for every fixed n, when p varies. This last fact implies, as Proposition 8.1.1 shows, that for a fixed n, the family $X(\mathrm{SL}_n(p); \{A_n, B_n\})$ has logarithmic diameter, but the proof is not effective, i.e., it does not give any hint how to represent a matrix in $\mathrm{SL}_n(p)$ as a short word in A_n and B_n. This is even open for $n = 2$.

Problem 10.8.3. *Does there exist a polynomial time algorithm (polynomial in $\log p$) which expresses an element of $\mathrm{SL}_2(p)$ (say $\begin{pmatrix} 1 & \frac{p-1}{2} \\ 0 & 1 \end{pmatrix}$) as a short word in $A_2 = \begin{pmatrix} 1 & 1 \\ 0 & 1 \end{pmatrix}$ and $B_2 = \begin{pmatrix} 0 & 1 \\ -1 & 0 \end{pmatrix}$?*

Note that if $p > 2$ and we allow to use a third generator $C = \begin{pmatrix} 2 & 0 \\ 0 & \frac{1}{2} \end{pmatrix}$, then this can be done quite easily. In fact the proof of Theorem 8.1.4 in [BKL] is effective.

A baby-version of Problem 10.8.3 is:

Problem 10.8.4. *Let p be an odd prime and $Y = \mathbb{F}_p \times \mathbb{F}_p \setminus \{(0,0)\}$. From $(a,b) \in Y$ we can move to either $(a \pm b, b)$ or $(a, b \pm a)$. Find a sequence of moves of length $O(\log p)$ which takes $(1,0)$ to $(\frac{p-1}{2}, 0)$.*

Note that Selberg's Theorem implies that such a sequence exists. This can be deduced from Theorem 4.4.2 by replacing the generator τ used there by $\begin{pmatrix} 1 & 0 \\ 1 & 1 \end{pmatrix}$. As mentioned there, Selberg's Theorem, whose proof is based on some spectral analysis on some Riemann surfaces plus the Riemann hypothesis for finite fields, is effective in the sense that the constant involved can be estimated. However, it gives no algorithm of how to find such a sequence of moves. It will be even more interesting if one can show that no such algorithm with $(\log p)^{O(1)}$ steps exists.

The results of §8.1 give a solution for the problem of finding «best case generators». L. Babai has conjectured (cf. [BHKLS]) that even «worst case generators» of finite non-abelian groups are not that bad:

Problem 10.8.5. *(L. Babai) Prove that there exists a constant C such that if G is a finite non-abelian simple group and S a set of generators of G then* $\mathrm{diam}X(G;S) \leq \log(|G|)^C$.

This is open even for the family of symmetric groups.

Very little is known about the diameters of Cayley graphs of a finite simple non-abelian group G with respect to k generators selected randomly. Recall that by Dixon [Dix] for the alternating groups, by Kantor-Lubotzky [KL] for the classical groups and by Liebeck-Shaler [LiS] for all the rest, almost every k-tuple (for $k \geq 2$) generate G. Problem 10.8.5 might be somewhat easier for random generators.

Problem 10.8.6. *For a fixed $k \geq 2$, prove that there exists a constant C such that $\mathrm{diam}X(G;S) \leq \log(|G|)^C$ for almost all k-tuples S of elements of G, where G is any finite non-abelian simple group, and almost all means except of a set whose proportion to $|G|^k$ is going to one when $|G|$ is going to infinity. Is it even true that* $\mathrm{diam}X(G;S) \leq C\log(|G|)$ *for almost all S?*

Some results in this direction for the symmetric groups were proved by Babai and Hetyei [BaHe]. Nothing is known on this issue for the group $\{SL_2(p) \mid p$ prime$\}$. This is a very interesting test case, though one should keep in mind that it might behave differently for the family $\{\mathrm{SL}_n(2) \mid n \geq 2\}$; see 10.3, [LW] and [GKS].

Passing to Section 8.2, we mention that the fact that we were able to compute so explicitly the eigenvalues of $X(\mathrm{SL}_2(p);S)$ from the character table of $\mathrm{SL}_2(p)$ is due to the fact that S was chosen to be a full conjugacy class.

Problem 10.8.7 *Find eigenvalues (or at least λ_1) of Cayley graphs with respect to some «nice generators», e.g., Coxeter groups with the standard generators,* $\mathrm{SL}_n(p)$ *with respect to elementary matrices, etc.*

Some results in this direction are given in [BDH] and [DG]. Roichman [Rn] obtained some powerful estimates on characters' values of the symmetric groups, which enabled him to estimate λ_1. Computational methods were developed in [LaRo] and the references therein. Finally we mention two problems extending the Ramanujan graphs in two directions:

Problem 10.8.8. (Ramanujan complexes) *Study the Laplacian of complexes (see [Ch], [Ga]), define Ramanujan complexes and construct such.*

Some work in this direction has been done for hyper-graphs ([FW], [Fr2] and [C3]).

Problem 10.8.9. Ramanujan diagrams:

(i) *Morgenstern ([Mo2], [Mo3]) studied the diagram associated with the standard characteristic p modular group. Study the diagrams obtained from different curves (see [S1]).*

(ii) *Prove Proposition 8.4.7 for $q = 3$ or 4. This will improve the known super-concentrators. It might be that once part (i) will be established, there will be other interesting finite subgraphs which might lead to some more applications to computer science.*

10.9 Distributing points on the sphere

Using the same notations as in Chapter 9, we denote by S a finite symmetric set of cardinality 2ℓ of isometries of a compact metric space X, e.g., $X = S^n$ the n-dimensional sphere, and $2\ell \cdot \lambda(S)$ is the norm of the operator $(\delta_S f)(x) = \sum_{\gamma \in S} f(\gamma x)$ acting on $L_0^2(X)$.

Problem 10.9.1. *Is there a fixed $\ell \in \mathbb{N}$ and $\varepsilon > 0$ such that for every $n \geq 2$, there exists a symmetric set $S(n)$ of 2ℓ elements of $SO(n+1)$ for which $\lambda(S) < 1 - \varepsilon$?*

This problem is a continuous analogue of Problem 10.3.4 above. If there exists a discrete group Γ with Property (T) with dense representation $\rho_n : \Gamma \to SO(n+1)$ for every n, then the problem has an affirmative solution by (the proof of) Proposition 9.1.3. However, all the known examples of discrete groups with Property (T) are lattices in some Lie groups. For all of them, super-rigidity results were proved which show that for a given Γ, such ρ_n exist for at most finitely many n's. The following is, however, open:

Problem 10.9.2. *Prove that $K = \prod_{n \geq 4} SO(n+1)$ contains no finitely generated dense subgroup with Property (T).*

A more detailed version of Problem 10.9.1 is:

Problem 10.9.3. *For n and ℓ, estimate from above and below the following quantity:*

$$\inf\{\lambda(S) \mid S \text{ a symmetric subset of } SO(n+1) \text{ of size } 2\ell \text{ acting on the } n-\text{ sphere}\}.$$

Problem 10.9.4. *What is $\lambda(S)$ for a random choice of a symmetric set of elements of $SO(n+1)$? Is it equal to 1 or smaller than 1? This is not known even for $n = 2$.*

Appendix: Modular forms, the Ramanujan conjecture and the Jacquet-Langlands correspondence

Jonathan D. Rogawski[1])

The theory developed in Chapter 7 relies on a fundamental result (Theorem 7.1.1) asserting that the space $L^2(\Gamma\backslash SO(3) \times PGL_2(\mathbb{Q}_p))$ decomposes as a direct sum of *tempered*, irreducible representations (see definition below). Here $SO(3)$ is the compact Lie group of 3×3 orthogonal matrices of determinant one, and Γ is a discrete group defined by a definite quaternion algebra D over $\mathbb{Q}$ which is split at p. The embedding of Γ in $SO(3) \times PGL_2(\mathbb{Q}_p)$ is defined by identifying $SO(3)$ and $PGL_2(\mathbb{Q}_p)$ with the groups of real and p-adic points of the projective group $D^*/\mathbb{Q}^*$.

Although this temperedness result can be viewed as a combinatorial statement about the action of the Hecke operators on the Bruhat-Tits tree associated to $PGL_2(\mathbb{Q}_p)$, it is not possible at present to prove it directly. Instead, it is deduced as a corollary of two other results. The first is the Ramanujan-Petersson conjecture for holomorphic modular forms, proved by P. Deligne [D]. The second is the Jacquet-Langlands correspondence for cuspidal representations of $GL(2)$ and multiplicative groups of quaternion algebras [JL]. The proofs of these two results involve essentially disjoint sets of techniques. Deligne's theorem is proved using the Riemann hypothesis for varieties over finite fields (also proved by Deligne) and thus relies on characteristic p algebraic geometry. By contrast, the Jacquet-Langlands Theorem is analytic in nature. The main tool in its proof is the Selberg trace formula. Furthermore, Deligne's Theorem is conveniently expressed in the classical language of modular forms, while the Jacquet-Langlands correspondence requires the language of infinite-dimensional representation theory. The purpose of this appendix is to explain the statements of the Deligne's Theorem and the Jacquet-Langlands correspondence, briefly outline their proofs, and explain how they combine to yield Theorem 7.1.1. The standard references [AA], [Co], [D3], [G] should be consulted for further details and related topics. I would like to thank Victor Tan for carefully reading the manuscript and suggesting several corrections.

---

1) Partially supported by a grant from the NSF.

A.0 Preliminaries

In this section, we fix notation and review some basic facts. See [CF] for a general treatment of adèles and idèles and [GPPS], [G] for a discussion of adèle groups.

Throughout this appendix, p will denote a prime of $\mathbb{Q}$ (the infinite prime $p = \infty$ is allowed). If p is finite, $\mathbb{Q}_p$ is the field of p-adic numbers and $\mathbb{Z}_p$ is the ring of p-adic integers. For $p = \infty$, $\mathbb{Q}_\infty = \mathbb{R}$. The absolute value on $\mathbb{Q}_p$ will be denoted by $|\ |_p$.

Let G denote the group $GL(2)$ of invertible 2×2 matrices, and define the subgroups:

$$B = \left\{ \begin{pmatrix} a & b \\ 0 & d \end{pmatrix} \in G \right\}, \quad M = \left\{ \begin{pmatrix} a & 0 \\ 0 & d \end{pmatrix} \in G \right\}, \quad N = \left\{ \begin{pmatrix} 1 & b \\ 0 & 1 \end{pmatrix} \in G \right\}.$$

The decomposition $B = MN$ holds. The center of G is the group of scalar matrices

$$Z = \left\{ \begin{pmatrix} a & 0 \\ 0 & a \end{pmatrix} \right\}.$$

We shall use algebraic group notation. Thus, if H is one of the above groups and R is any ring, then $H(R)$ is the group of R-valued points of H, i.e., the group of elements of H with entries from R. We will write H_p for $H(\mathbb{Q}_p)$.

For all p, G_p is a locally compact group under the topology inherited from the natural coordinate topology on the set of 2×2 matrices. For $p = \infty$, G_∞ is a Lie group. For $p < \infty$, G_p is totally disconnected, in the sense that there exists a basis for the open neighborhoods of the identity consisting of subgroups. Such a basis is given by the set of congruence subgroups $1 + p^n M_2(\mathbb{Z}_p)$ for $n \geq 1$.

Let K_p be the following standard maximal compact subgroup of G_p. If $p < \infty$, $K_p = GL(\mathbb{Z}_p)$. For $p = \infty$, let

$$r(\theta) = \begin{pmatrix} \cos\theta & \sin\theta \\ -\sin\theta & \cos\theta \end{pmatrix}$$

and set

$$K_\infty = \{ r(\theta) : \theta \in [0, 2\pi) \}.$$

The *Iwasawa decomposition*

$$G_p = B_p K_p$$

holds for all p.

As explained in Chapter 6, the *adèle ring* $\mathbb{A}$ of $\mathbb{Q}$ is the restricted direct product of the fields $\mathbb{Q}_p$:

$$\mathbb{A} = \left\{ (a_p) \in \prod_p \mathbb{Q}_p : a_p \in \mathbb{Z}_p \text{ for almost all primes } p < \infty \right\}.$$

Let $\mathbb{A}_f$ be the ring of finite adèles, defined as a restricted direct product as above but without the infinite factor $\mathbb{R}$. When convenient, we also identify $\mathbb{A}_f$ with the subring of elements $(a_p) \in \mathbb{A}$ such that $a_\infty = 0$. For any adèle a, we write a_f for the finite part of a, i.e., the tuple obtained by dropping a_∞.

We view $\mathbb{Q}$ as a subring of $\mathbb{A}$ relative to the diagonal embedding. Thus $x \in \mathbb{Q}$ is identified with the adèle $(x, x, \ldots)$. Let

$$\widehat{\mathbb{Z}} = \varprojlim_N \mathbb{Z}/N = \prod_{p<\infty} \mathbb{Z}_p.$$

The ring $\mathbb{A}$ has the structure of a locally compact topological ring for the topology relative to which $\mathbb{R} \times \widehat{\mathbb{Z}}$ with its product topology is an open subring (cf. 6.1). The image of $\mathbb{Q}$ in $\mathbb{A}$ is a discrete subgroup since $\mathbb{Q} \cap (-1,1) \times \widehat{\mathbb{Z}} = \{0\}$. Furthermore, there is a surjective homomorphism of $\widehat{\mathbb{Z}}$ onto $\mathbb{Z}/N$ with kernel $N\widehat{\mathbb{Z}}$, and for all integers N we have the following decomposition (referred to as the Strong Approximation Theorem):

$$\mathbb{A} = \mathbb{Q} + (\mathbb{R} \times N\widehat{\mathbb{Z}}).$$

It follows that $\mathbb{Q}\backslash\mathbb{A} = N\mathbb{Z}\backslash(\mathbb{R} \times N\widehat{\mathbb{Z}})$ for all N and also that $\mathbb{Q}\backslash\mathbb{A}$ is compact.

The *idèle group* $\mathbb{I}$ is the group of invertible elements in $\mathbb{A}$:

$$\mathbb{I} = \left\{ (a_p) \in \prod_p \mathbb{Q}_p^* : a_p \in \mathbb{Z}_p^* \text{ for almost all primes } p < \infty \right\}.$$

The group $\mathbb{I}_f$ of finite idèles is the group of invertible elements in $\mathbb{A}_f$. Note that $\widehat{\mathbb{Z}}^* = \prod_{p<\infty} \mathbb{Z}_p^*$. The idèle group is a locally compact group for the topology relative to which $\mathbb{R}^* \times \widehat{\mathbb{Z}}^*$ with its product topology is an open subgroup. The image of $\mathbb{Q}^*$ in $\mathbb{I}$ is a discrete subgroup since $\mathbb{Q}^* \cap \mathbb{R}_+^* \times \widehat{\mathbb{Z}}^* = \{1\}$. Here $\mathbb{R}_+^*$ is the multiplicative group of positive real numbers and $\mathbb{Q}^*$ is identified with its image under the diagonal embedding. For $x \in \mathbb{I}$, the absolute value is defined as the product of the local absolute values:

$$|x| = \prod_p |x|_p.$$

The product is actually finite since $|x|_p = 1$ for almost all p. Let $\mathbb{I}^1$ be the group of idèles x such that $|x| = 1$. The product formula asserts that $\mathbb{Q}^*$ is contained in $\mathbb{I}^1$. It follows from the unique factorization into primes that $\mathbb{I}$ decomposes as a direct product

$$\mathbb{I} = \mathbb{Q}^* \left(\mathbb{R}_+^* \times \widehat{\mathbb{Z}}^* \right). \tag{1.1}$$

In fact, for any idèle x, define the rational number $\alpha = \mathrm{sgn}(x_\infty) \prod_{p<\infty} |x_p|$. Then $\alpha x \in \mathbb{R}_+^* \times \widehat{\mathbb{Z}}^*$.

The adèle group $G(\mathbb{A})$ is the group of invertible 2×2 matrices with entries from $\mathbb{A}$:

$$G(\mathbb{A}) = \left\{ g = (g_p) \in \prod_p G(\mathbb{Q}_p) : g_p \in G(\mathbb{Z}_p) \text{ for almost all } p \right\}.$$

This is a locally compact group for the topology relative to which $G(\mathbb{R}) \times \prod_{p<\infty} G(\mathbb{Z}_p)$ with its product topology is an open subgroup.

A *Hecke character* ω is a continuous character of $\mathbb{I}$ which is trivial on $\mathbb{Q}^*$. Let ω_p and ω_f denote the restrictions of ω to $\mathbb{Q}_p^*$ and I_f, respectively. By (1.1), we may view ω as a continuous character of $\mathbb{R}_+^* \times \widehat{\mathbb{Z}}^*$. For all $s \in \mathbb{C}$, $x \to |x|^s$ is a Hecke character which is trivial on $\widehat{\mathbb{Z}}^*$. On the other hand, if ω' is a Dirichlet character, i.e., a character of $(\mathbb{Z}/N)^*$, then ω' defines a Hecke character ω which is trivial on $\mathbb{R}_+^*$ via the homomorphism «reduction modulo N»:

$$\widehat{\mathbb{Z}}^* \to (\mathbb{Z}/N)^*.$$

Observe that with this definition, if $(p, N) = 1$, then $\omega_p(p) = \omega'(p)^{-1}$. Indeed,

$$\begin{aligned} 1 = \omega((p, p, p, \ldots)) &= \omega_p(p)\omega((p, p, \ldots, 1, p, \ldots)) \\ &= \omega_p(p)\omega'(p), \end{aligned}$$

where $\alpha = (p, p, \ldots, 1, p, \ldots)$ is the idèle such that $\alpha_p = 1$ and $\alpha_\ell = p$ for all primes $\ell \neq p$. Every Hecke character ω decomposes uniquely as a product $\omega_0(x)|x|^s$, where ω_0 is defined by a Dirichlet character. Note that $Z(\mathbb{A})$ is isomorphic to $\mathbb{I}$. We may therefore regard any Hecke character as a character of $Z(\mathbb{Q})\backslash Z(\mathbb{A})$.

A.1 Representation theory and modular forms

A *unitary representation* of a topological group H is a pair (π, V), where V is a Hilbert space and π is a homomorphism from H to the group $U(H)$ such that the map

$$H \times V \to V$$

$$(h, v) \to \pi(h)v$$

is continuous. Sometimes, we write V_π for the space V, and at other time, we drop V from the notation and notate only the homomorphism π.

If H is a unimodular locally compact group and H' is a closed unimodular subgroup of H, then there exists an H-invariant measure on the quotient space $H'\backslash H$ and we may define the Hilbert space $L^2(H'\backslash H)$. The group H acts unitarily on $L^2(H'\backslash H)$ by right translation ρ:

$$(\rho(h)\phi)(x) = \phi(xh).$$

In the theory of automorphic forms, we are interested in the representation of $G(\mathbb{A})$ on $L^2(G(\mathbb{Q})\backslash G(\mathbb{A}))$. However, the quotient $Z(\mathbb{Q})\backslash Z(\mathbb{A})$ is noncompact, and it is more convenient to consider a slightly different Hilbert space. Fix a unitary character

$$\omega : Z(\mathbb{Q})\backslash Z(\mathbb{A}) \to \mathbb{C}^*$$

and let $L^2(\omega)$ be the Hilbert space of functions ϕ on $G(\mathbb{Q})\backslash G(\mathbb{A})$ such that $\phi(zg) = \omega(z)\phi(g)$ and

$$\int_{G(\mathbb{Q})Z(\mathbb{A})\backslash G(\mathbb{A})} |\phi(g)|^2 \, d\dot{g} < \infty$$

where $d\dot{g}$ is a $G(\mathbb{A})$-invariant measure on $G(\mathbb{Q})Z(\mathbb{A})\backslash G(\mathbb{A})$. Let ρ_ω be the representation on $L^2(\omega)$ by right translation. We may assume, without loss of generality, that ω is a Dirichlet character. Indeed, suppose that $\omega' = \omega|\ |^s$, and let χ_s be the character $\chi_s(g) = |\det(g)|^{\frac{s}{2}}$. Then $\rho_\omega \otimes \chi_s$ is isomorphic to $\rho_{\omega'}$ via the map sending $\phi(g) \in L^2(\omega)$ to $\phi(g)\chi_s(g)$ in $L^2(\omega')$.

The space of cusp forms

The *constant term* of a function $\phi \in L^2(\omega)$ is defined by

$$\phi_N(g) = \int_{N(\mathbb{Q})\backslash N(\mathbb{A})} \phi(ng) \, dn.$$

The quotient $N(\mathbb{Q})\backslash N(\mathbb{A})$ is isomorphic to $\mathbb{Q}\backslash\mathbb{A}$, hence compact, and thus the integral is defined for almost all g. The function ϕ is said to be *cuspidal* if $\phi_N(g) = 0$ for almost all $g \in G(\mathbb{A})$. Since this condition is invariant under the right action of $G(\mathbb{A})$, the subspace $L_0^2(\omega)$ of all cuspidal functions is a closed $G(\mathbb{A})$-invariant subspace of $L^2(\omega)$.

Denote the restriction of ρ_ω to $L_0^2(\omega)$ by $\rho_{0\omega}$. By a fundamental theorem of Gelfand and Piatetski-Shapiro ([GGPS], [La]), $\rho_{0\omega}$ is completely reducible, that is

$$\rho_{0\omega} \xrightarrow{\sim} \bigoplus m(\pi)\pi.$$

Here π ranges through a countable set of inequivalent irreducible unitary representations of G, and the multiplicity $m(\pi)$ of π is finite. The representations occurring with nonzero multiplicity in $L_0^2(\omega)$ for some ω are called *cuspidal representations*. The cuspidal representations are the primary object of study in the theory of automorphic forms.

The following basic result is due to Jacquet and Langlands [JL].

Theorem 1.1. *If (π, V) is cuspidal, then $m(\pi) = 1$.*

In other words, if (π, V) is an irreducible unitary representation of $G(\mathbb{A})$, then there is at most one isometric intertwining operator $T : V \to L_0^2(\omega)$. Theorem 1.1 is a consequence of a result in the representation theory of the local groups G_p, the uniqueness of the so-called Whittaker models (cf. article of I. Piatetski-Shapiro in [Co], Vol. 1).

Congruence subgroups

If $\Gamma \subset SL_2(\mathbb{Z})$ is a subgroup of finite index, we can consider the representation of $SL_2(\mathbb{R})$ by right translation on the Hilbert space $L^2(\Gamma\backslash SL_2(\mathbb{R}))$. Let

$$\Gamma(N) = \ker\left(SL_2(\mathbb{Z}) \to SL_2(\mathbb{Z}/N)\right),$$

where the arrow is reduction modulo N. Recall that a subgroup $\Gamma \subset SL_2(\mathbb{Z})$ is said to be a *congruence subgroup* if it contains $\Gamma(N)$ for some N. The spaces $L^2(\Gamma\backslash SL_2(\mathbb{R}))$ for Γ a congruence subgroup can be analyzed in terms of the adelic spaces $L^2(\omega)$. We shall illustrate with the Hecke subgroups $\Gamma_0(N)$.

Let

$$\Gamma_0(N) = \left\{\begin{pmatrix} a & b \\ c & d \end{pmatrix} \in SL_2(\mathbb{Z}) : c \equiv 0 \bmod N\right\}$$

and set

$$\Gamma_1(N) = \left\{\begin{pmatrix} a & b \\ c & d \end{pmatrix} \in \Gamma_0(N) : a \equiv d \equiv 1 \bmod N\right\}.$$

Then $\Gamma_1(N)$ is a normal subgroup of $\Gamma_0(N)$ and the quotient $\Gamma_1(N)\backslash\Gamma_0(N)$ acts on

$$L^2(N) = L^2(\Gamma_1(N)\backslash SL_2(\mathbb{R})),$$

with $\gamma \in \Gamma_0(N)$ acting by $f(g) \to f(\gamma^{-1}g)$. Since $\Gamma_1(N)\backslash\Gamma_0(N)$ is isomorphic to $(\mathbb{Z}/N)^*$, a Dirichlet character ω' of $(\mathbb{Z}/N)^*$ may be viewed as a character of $\Gamma_0(N)$ by the formula

$$\omega'\left(\begin{pmatrix} a & b \\ c & d \end{pmatrix}\right) = \omega'(d).$$

Let $L^2_{\omega'}(N)$ be the ω'-eigenspace of $\Gamma_0(N)$ acting on $L^2(N)$. As described in 0, ω' defines a Hecke character ω, which we also view as a character of $Z(\mathbb{Q})\backslash Z(\mathbb{A})$. The group $SL_2(\mathbb{R})$ acts on $L^2(\omega)$ since it is a subgroup of G_∞. We now define an embedding

$$L^2_{\omega'}(N) \to L^2(\omega),$$

which is equivariant for the action of $SL_2(\mathbb{R})$ on both spaces. Set

$$K = G(\widehat{\mathbb{Z}}) = \prod_{p<\infty} K_p.$$

Reduction modulo N defines a surjective homomorphism

$$K \to GL_2(\mathbb{Z}/N).$$

Let $K_0(N)$ be the subgroup of elements $k \in K$ whose reduction modulo N is upper-triangular. The following decomposition holds

$$G(\mathbb{A}) = G(\mathbb{Q})\left(G^+_\infty \times K_0(N)\right) \tag{1.2}$$

where G^+_∞ is the subgroup of elements with positive determinant [G]. Note that $G^+_\infty = Z(\mathbb{R})^+ SL_2(\mathbb{R})$ where $Z(\mathbb{R})^+$ is the subgroup of $Z(\mathbb{R})$ of scalar matrices with positive diagonal entries. For

$$k_f = \begin{pmatrix} a & b \\ c & d \end{pmatrix} \in K_0(N),$$

set $\omega_f(k_f) = \omega'(d')$ where d' is the image of d in $(\mathbb{Z}/N)^*$. For $f \in L^2_{\omega'}(N)$, define

$$\phi_f(\gamma(z_\infty g_\infty \times k_f)) = \omega_f(k_f) f(g_\infty)$$

where $\gamma \in G(\mathbb{Q})$, $z_\infty \in Z(\mathbb{R})^+$, $g_\infty \in SL_2(\mathbb{R})$, and $k_f \in K_0(N)$. It follows from the equality

$$\Gamma_0(N) = G(\mathbb{Q}) \cap (G^+_\infty \times K_0(N)) \tag{1.3}$$

that ϕ_f is well-defined. Indeed, if $\delta \in \Gamma_0(N)$, then

$$\begin{aligned}\phi_f(\gamma\delta(z_\infty\delta_\infty^{-1}g_\infty \times \delta_f^{-1}k_f) &= \omega_f(\delta_f^{-1}k_f)f(\delta^{-1}g_\infty)\\ &= \omega_f(k_f)f(g_\infty)\\ &= \phi_f(z_\infty g_\infty \times k_f).\end{aligned}$$

The map $f \to \phi_f$ defines the $SL_2(\mathbb{R})$-equivariant embedding $L^2_\omega(N) \to L^2(\omega)$.

If N divides M, we may also view ω' as character of $(\mathbb{Z}/M)^*$, and there is a natural inclusion

$$L^2_{\omega'}(N) \to L^2_{\omega'}(M)$$

compatible with the inclusions of both spaces into $L^2(\omega)$. The space $L^2(\omega)$ is the closure of the union of the images of the spaces $L^2_{\omega'}(N)$. The advantage of $L^2(\omega)$ over the spaces $L^2_{\omega'}(N)$ is that the former Hilbert space is equipped with an action of the large group $G(\mathbb{A})$, whereas $L^2_{\omega'}(N)$ is a representation space only for $SL_2(\mathbb{R})$.

A function $f \in L^2_{\omega'}(N)$ is said to be cuspidal if for all $\delta \in SL_2(\mathbb{Z})$,

$$\int_{N(\mathbb{R})\cap\delta^{-1}\Gamma_1(N)\delta\backslash N(\mathbb{R})} f(\delta n g)\, dn = 0$$

for almost all $g \in SL_2(\mathbb{R})$. The space of cuspidal functions $L^2_{0\omega'}(N)$ maps to $L^2_0(\omega)$. The theorem of Gelfand and Piatetski-Shapiro also holds in this setting:

$$L^2_{0\omega'}(N) \xrightarrow{\sim} \bigoplus m(\pi)\pi.$$

Here π ranges through a countable set of inequivalent irreducible unitary representations of $SL_2(\mathbb{R})$. The multiplicities $m(\pi)$ are finite, but not necessarily equal to one. The multiplicity one theorem (Theorem 1.1 above) holds only for representations of the big group $G(\mathbb{A})$.

Modular forms

Now we will show how classical modular forms fit into the above framework. A (classical) holomorphic modular form of weight k with respect to $\Gamma_0(N)$ and Dirichlet character ω' of $\mathbb{Z}/N$ (also called the *nebentypus*) is an analytic function $f(z)$ on the upper half-plane

$$\mathbb{H} = \{z = x + iy \in \mathbb{C} : y > 0\}$$

which satisfies two conditions. First, the transformation law of weight k:

$$f(\gamma(z)) = \omega'(d)^{-1} j(\gamma, z)^k f(z) \quad \text{for all} \quad \gamma \in \Gamma_0(N). \tag{1.4}$$

Here $j(\gamma, z) = cz + d$ is the so-called *factor of automorphy* and

$$\gamma(z) = \frac{az+b}{cz+d} \quad \text{for} \quad \gamma = \begin{pmatrix} a & b \\ c & d \end{pmatrix}.$$

We need to assume that $\omega'(-1) = (-1)^k$; otherwise (1.4) forces f to be 0. The second condition is the requirement of *regularity at* ∞ defined as follows. For $\delta \in SL_2(\mathbb{Q})$, define

$$f_\delta(z) = f(\delta(z)) j(\delta, z)^{-k}.$$

The cocycle relation

$$j(gh, z) = j(g, h(z)) j(h, z)$$

is easily verified, and it implies that f_δ satisfies the weight k transformation law (1.4) for the group $\delta^{-1}\Gamma_0(N)\delta$. There exists a positive integer M such that

$$N(\mathbb{Q}) \cap \delta^{-1}\Gamma_0(N)\delta = \left\{ \begin{pmatrix} 1 & tM \\ 0 & 1 \end{pmatrix} : t \in \mathbb{Z} \right\},$$

and hence $f_\delta(z+M) = f_\delta(z)$. Because of this periodicity, f may be viewed as a function of the variable $q = e^{2\pi i z/M}$. The transformation $z \to q$ maps $\mathbb{H}$ onto the punctured unit disc $\Delta^* = \{z \in \mathbb{C} : 0 < |z| < 1\}$. Hence f_δ defines an analytic function on Δ^*, and as such, it has a Laurent expansion in the variable q,

$$f_\delta(z) = \sum_{n \in \mathbb{Z}} a_n q^n = \sum_{n \in \mathbb{Z}} a_n e^{2\pi i n z/M}.$$

The numbers a_n are called the Fourier coefficients of f_δ. We say that f is regular at infinity if for all δ, the Fourier coefficients of f_δ satisfy $a_n = 0$ for $n < 0$. In other words, $f_\delta(q)$ is analytic at $q = 0$. If, in addition, $a_0 = 0$ for all δ, then f is said to be a *cusp* form.

Let $S_k(N, \omega')$ be the space of cusp forms of weight k and nebentypus ω', and let ω be the Hecke character defined by ω'. For $f \in S_k(N, \omega')$, define a function ϕ_f on $G(\mathbb{A})$ by

$$\phi_f(\gamma(z_\infty g_\infty \times k_f) = \omega_f(k_f) j(g_\infty, i)^{-k} f(g_\infty(i)) \tag{1.5}$$

where $\gamma \in G(\mathbb{Q})$, $z_\infty \in Z(\mathbb{R})^+$, $g_\infty \in SL_2(\mathbb{R})$, and $k_f \in K_0(N)$. It follows from (1.3) and the cocycle relation that ϕ_f is a well-defined function on $G(\mathbb{Q})\backslash G(\mathbb{A})$ and satisfies $\phi_f(zg) = \omega(z)\phi_f(g)$ for $z \in \mathbb{Z}(\mathbb{A})$.

We now show that the cuspidality of f implies that ϕ_f is cuspidal. Let $\phi = \phi_f$ and let $g = g_\infty \times g_f \in G(\mathbb{A})$. Let $z = g_\infty(i)$. By (1.2), $G(\mathbb{A}_f) = G(\mathbb{Q})K_0(N)$ and hence there exists $\delta \in G(\mathbb{Q})$ such that $k_f = \delta_f g_f$ lies in $K_0(N)$. Then

$$\begin{aligned} \phi_f(g_\infty \times g_f) &= \phi_f(g_\infty \times \delta_f^{-1} k_f) = \phi_f(\delta_\infty g_\infty \times k_f) \\ &= \omega_f(k_f) j(\delta g_\infty, i)^{-k} f(\delta(z)) \\ &= \omega_f(k_f) j(g_\infty, i)^{-k} f_\delta(z). \end{aligned}$$

For some $M \in \mathbb{Z}$, we have

$$N'' = N(\mathbb{Q}) \cap \delta^{-1}\Gamma_0(N)\delta = \left\{ \begin{pmatrix} 1 & tM \\ 0 & 1 \end{pmatrix} : t \in \mathbb{Z} \right\}.$$

For $n \in N(\mathbb{A})$ and $n' \in N(M\widehat{\mathbb{Z}})$, we have

$$\phi(nn'g) = \phi(ng(g_f^{-1} n' g_f) = \phi(ng)$$

since $g_f^{-1}n'g_f = k_f^{-1}\delta_f n'\delta_f^{-1}k_f$ belongs to $K_0(N)$. By strong approximation, $N(\mathbb{Q})\backslash N(\mathbb{A}) = N''\backslash(N(\mathbb{R}) \times N(M\widehat{\mathbb{Z}}))$, and thus

$$\begin{aligned}\phi_N(g) &= \int_{N(\mathbb{Q})\backslash N(\mathbb{A})} \phi(ng)dn = \int_{N''\backslash N(\mathbb{R})} \phi(n_\infty g)dn_\infty \\ &= \omega_f(k_f)j(g_\infty, i)^{-k} \int_0^M f_\delta(z+t)dt.\end{aligned}$$

This vanishes if f_δ is cuspidal.

The function ϕ_f satisfies an additional transformation property. The subgroup K_∞ is the stabilizer in G of the point $i \in \mathbb{H}$. The relation

$$j(\gamma r(\theta), i) = e^{-i\theta} j(\gamma, i),$$

together with (1.5) and the invariance of ϕ_f under $Z(\mathbb{R})$, implies

$$\phi_f(g \begin{pmatrix} a & b \\ -b & a \end{pmatrix}_\infty) = \left(\frac{a+bi}{|a+bi|}\right)^k \phi_f(g). \tag{1.6}$$

The cuspidality of f also implies that ϕ is square-integrable. Observe that $|\phi(g)|^2$ is right $K_\infty K_0(N)$-invariant. By (1.2),

$$\begin{aligned}\int_{G(\mathbb{Q})Z(\mathbb{A})\backslash G(\mathbb{A})} |\phi(g)|^2 \, d\dot{g} &= \int_{\Gamma_0(N)\backslash SL_2(\mathbb{R})/K_\infty} |\phi(g_\infty)|^2 \, d\dot{g_\infty} \\ &= \int_{\Gamma_0(N)\backslash SL_2(\mathbb{R})/K_\infty} |j(g_\infty, i)^{-k} f(g_\infty(i))|^2 \, d\dot{g_\infty}\end{aligned}$$

for suitable normalization of measures. The G_∞-invariant measure on $\mathbb{H} = SL_2(\mathbb{R})/K_\infty$ is $y^{-2}dxdy$. Since $j(g_\infty, i) = \operatorname{Im}(g_\infty(i))^{-\frac{1}{2}}$, we see that

$$\int_{G(\mathbb{Q})Z(\mathbb{A})\backslash G(\mathbb{A})} |\phi(g)|^2 \, d\dot{g} = \int_{\Gamma_0(N)\backslash\mathbb{H}} |f(x+iy)|^2 y^k \frac{dx\,dy}{y^2}.$$

An elementary estimate [Sh1] shows that for all $\delta \in SL_2(\mathbb{Q})$, there exists $C > 0$ such that $|f_\delta(x+iy)| < Ce^{-y}$ as $y \to \infty$. By the well-known structure of fundamental domains for $\Gamma_0(N)$, this easily implies that $\phi_f(g)$ is square-integrable.

Examples of modular forms

1. Let $k \geq 4$ and set

$$E_k(z) = \sum_{(m,n)\neq(0,0)} (mz+n)^{-k}.$$

Here (m,n) ranges over all integer pairs other than $(0,0)$. The sum is absolutely convergent, and convergence is uniform for z in a compact subset of $\mathbb{H}$. Hence $E_k(z)$ is an analytic function of z. Direct calculation shows that E_k satisfies the weight k transformation law (1.4) for the trivial character ω' and the group $SL_2(\mathbb{Z})$. It is not difficult to verify regularity at ∞ [Se].

2. Let S_k be the space of cusp forms for the group $SL_2(\mathbb{Z})$ and the trivial character ω' . As shown in [Se], $S_k = \{0\}$ for $k \leq 10$. However, S_{12} is one-dimensional, spanned by the cusp form

$$\Delta(z) = (2\pi)^{-12}(60E_4(z))^3 - 27(140E_3(z))^4 = \sum_{n\geq 1} \tau(n)q^n.$$

The second expression is the Fourier expansion of Δ with respect to the parameter $q = e^{2\pi i z}$. The Fourier coefficients $\tau(n)$ define the so-called Ramanujan tau function. There is another well-known formula for Δ:

$$\Delta(z) = q\prod_{n\geq 1}(1-q^n) = q - 24q + 252q^3 + \cdots.$$

The Ramanujan Conjectures

Ramanujan conjectured two fundamental properties of the tau function:

(1) $\tau(n)$ satisfies the multiplicative relations

$$\tau(p^{n+1}) = \tau(p)\tau(p^n) - p^{11}\tau(p^{n-1}) \quad \text{for } p \text{ a prime,}$$
$$\tau(n)\tau(m) = \tau(mn) \quad \text{if} \quad (m,n) = 1;$$

(2) for all primes p, $|\tau(p)| \leq 2p^{\frac{11}{2}}$.

The first conjecture is essentially elementary. It was proved by Mordell using a special case of the so-called Hecke operators.

We recall the definition of the Hecke operators and some basic facts (cf. [Sh1], and also [Se]). Let p be a prime. For $f \in S_k(N,\omega')$, define

$$T_p f(z) = \omega'(p)^{-1}p^{k-1}f(pz) + p^{-1}\sum_{j=0}^{p-1} f(\frac{z+j}{p}),$$

where the first term does not appear if p divides N. Then $T_p f$ again belongs to $S_k(N,\omega')$, and T_p defines a linear operator on $S_k(N,\omega')$. The operators T_p for different primes commute with each other. Hence, there exists a basis of $S_k(N,\omega')$ relative to which the operators T_p act by upper-triangular matrices. It can be shown that T_p is diagonalizable for $(p,N) = 1$. A *Hecke eigenform* is a simultaneous eigenfunction f for all of the operators T_p.

Let f be a Hecke eigenform and consider the Fourier expansion of f relative to the cusp ∞, i.e. the expansion of $f(z)$ in powers of $q = e^{2\pi iz}$. It is known that the first Fourier coefficient a_1 is necessarily nonzero. We say that f is normalized if $a_1 = 1$. Assuming f normalized, we have

$$f(z) = q + \sum_{n>1} a_n q^n,$$

and a simple calculation shows that the n-th Fourier coefficient a'_n of $T_p f$ is given by the formula

$$a'_n = a_{np} + \omega'(p)^{-1} p^{k-1} a_{n/p} \tag{1.7}$$

(where $a_{n/p} = 0$ if p does not divide n and the second term on the right is zero if p divides N). In particular, $a'_1 = a_p$ and thus

$$a_p = \lambda_p$$

where λ_p is the eigenvalue of f relative to T_p, i.e., $T_p f = \lambda_p f$. This yields the recursion relation

$$a_p a_n = a_{np} + \omega'(p)^{-1} p^{k-1} a_{n/p}. \tag{1.8}$$

It is straightforward to check that Ramanujan's multiplicative relations for $\tau(n)$ follow from (1.8) and hence are *equivalent* to the assertion that Δ is a Hecke eigenform. This latter assertion is an immediate consequence of the fact that the space of weight 12 cusp forms on $SL_2(\mathbb{Z})$ is one-dimensional [Se].

The Ramanujan-Petersson Conjecture

The natural generalization of Ramanujan's conjecture on the size of $\tau(p)$ is the so-called Ramanujan-Petersson (RP) conjecture. It asserts that if $f \in S_k(N, \omega')$ is a Hecke eigenform of weight k, then the eigenvalues λ_p satisfy the inequality $|\lambda_p| \leq 2p^{\frac{k-1}{2}}$ for $(p, N) = 1$. This was first proved in the special case of weight $k = 2$. In this case, the inequality $|\lambda_p| \leq 2p^{\frac{1}{2}}$ follows from the Eichler-Shimura theory and the Riemann hypothesis for curves over finite fields.

We briefly recall the statement of the Riemann hypothesis for curves. Suppose that X is a smooth projective curve of genus g over the finite field $\mathbb{F}_q$. Then X is defined as the zero set in projective space $\mathbb{P}^m$ of a collection of homogeneous polynomials $\{F_j(x_1, \ldots, x_{m+1})\}$ in $(m+1)$ variables. Let N_n be the number of points on X defined over $\mathbb{F}_{q^n}$, i.e., the number of solutions to the system $F_j(x_1, \ldots, x_{m+1}) = 0$ in projective space over $\mathbb{F}_{q^n}$. It was first shown by F.K. Schmidt (1931) that N_n is given by a formula of a particularly simple type: there exist complex numbers $\alpha_1, \ldots, \alpha_{2g}$ such that

$$N_n = 1 + q^n - \sum_{j=1}^{2g} \alpha_j^n \tag{1.9}$$

for all $n \geq 1$. This can be expressed as a statement about the rationality of the zeta function attached to X, defined by

$$Z(T, X_{/\mathbb{F}_q}) = \exp\left(\sum_{n\geq 1} \frac{N_n}{n} T^n\right).$$

Equality (1.9) is equivalent to the formula

$$Z(T, X_{/\mathbb{F}_q}) = \frac{\prod_{j=1}^{2g}(1 - \alpha_j T)}{(1 - T)(1 - qT)}.$$

The zeta function also satisfies a functional equation, which is equivalent to the assertion that the set of numbers α_j is invariant under the transformation $\alpha_j \to q/\alpha_j$. The *Riemann hypothesis* (RH), proved by A. Weil, asserts that $|\alpha_j| = q^{1/2}$. The name «Riemann hypothesis» is used in this context because the numbers α_j^{-1} are the zeroes of $Z(T, X)$ and hence the zeroes of the analytic function $Z(q^{-s}, X)$ have real part $\mathrm{Re}(s) = \frac{1}{2}$.

In 1954, Eichler [E] discovered the connection between the RP conjecture and the Riemann hypothesis. Let α_p, β_p be the roots of the polynomial

$$X^2 - \lambda_p X + \omega'(p)^{-1} p.$$

Then the equality $|\lambda_p| \leq 2p^{\frac{1}{2}}$ is equivalent to $|\alpha_p| = |\beta_p| = p^{\frac{1}{2}}$ (since $|\alpha_p \beta_p| = p$). The RP conjecture in weight 2 would follow from Weil's Theorem (the Riemann hypothesis) if one could show that there is a curve over $\mathbb{F}_p$ such that α_p and β_p appear among the α_j's in (1.9). Eichler showed that in the case of trivial ω, the appropriate curve is the modular curve $X_0(N)$ associated to $\Gamma_0(N)$. This is a projective algebraic curve defined over $\mathbb{Q}$ whose set of complex points is isomorphic to the compactification of the Riemann surface $\Gamma_0(N)\backslash\mathbb{H}$. Since $X_0(N)$ is defined over $\mathbb{Q}$, the reduction modulo p of its defining equations yields a smooth projective algebraic curve over $\mathbb{F}_p$ for all p outside a finite set S. Eichler showed that for $p \notin S$, the set of the inverse zeroes of the zeta function $Z(T, X_0(N)_{/\mathbb{F}_p})$ coincides with the set of numbers α_p, β_p arising from weight two modular forms as above. Subsequently, Igusa showed that S is the set of primes dividing N. More general congruence subgroups were treated by Shimura ([Sh1], [D1], [Kn2]).

The proof of RP for weights $k > 2$ is based on the Riemann hypothesis for higher-dimensional varieties. Suppose that $X_{/\mathbb{Q}}$ is a smooth projective variety over $\mathbb{Q}$ of dimension M with good reduction at p (i.e., when the equations defining X are reduced modulo p, they define a smooth projective variety over $\mathbb{F}_p$). As before, let N_n be the number of points on X defined over $\mathbb{F}_{q^n}$. Let $b(j)$ be the jth Betti number of $X(\mathbb{C})$, i.e., the dimension of $H^j(X(\mathbb{C}), \mathbb{R})$. As proved by

B. Dwork (1960) and A. Grothendieck, for $0 \le i \le 2M$, there exist complex numbers $\alpha_{i1}, \alpha_{i2}, \ldots, \alpha_{ib(i)}$ such that for all $n \ge 1$,

$$N_n = \sum (-1)^j \alpha_{ij}^n.$$

The numbers α_{ij} are the eigenvalues of the Frobenius endomorphism acting on the ℓ-adic cohomology group $H_\ell^i(X)$. The Riemann hypothesis for varieties over finite fields is the statement

$$|\alpha_{ij}| = q^{\frac{i}{2}}.$$

We refer to [Ka] for an excellent discussion of Deligne's proof of the RH.

The RP conjecture for Hecke eigenforms of weight k can be formulated as the statement that the roots α_p, β_p of

$$X^2 - \lambda_p X + \omega'(p)^{-1} p^{k-1} \tag{1.10}$$

satisfy $|\alpha_p| = |\beta_p| = p^{\frac{k-1}{2}}$. This would follow if one could show that α_p, β_p occur as Frobenius eigenvalues in $H_\ell^{k-1}(X)$ for some smooth projective variety X. For $k = 3$, the natural variety to consider is the 2-dimensional fiber system of elliptic curves over $\Gamma_0(N)\backslash\mathbb{H}$. For $k \ge 3$, one uses the $(k-2)$-nd symmetric power of this fiber system. This approach was suggested by Kuga and Shimura. The substantial obstacles to carrying it out, arising from the noncompactness of the fiber systems, were overcome by Deligne [D1].

Theorem 1.2. (Deligne) *Let $f \in S_k(N, \omega')$ and suppose $T_p f = \lambda_p f$ for $(p, N) = 1$. Then $|\lambda_p| \le 2p^{\frac{k-1}{2}}$.*

The Eichler-Shimura theory, giving the relation between zeta functions and modular forms, has undergone a significant development in the past thirty years. During the 1960s, Shimura created the theory of canonical models, thus laying the arithmetic foundations for a general theory of zeta functions of the so-called *Shimura varieties* attached to reductive groups. In the 1970s, Langlands introduced the machinery of infinite-dimensional representation theory to formulate general conjectures and a program for proving them using the trace formula. We refer to [BR] and the article of R. Kottwitz in [AA] for further information about these developments.

A.2 Classification of unitary representations

In this section, we review the classification of irreducible unitary representations of G_p. Our motivation comes from the fact that the irreducible unitary representations of $G(\mathbb{A})$ can be constructed as suitable tensor products $\otimes\pi_p$ of representations of the local groups G_p (see below).

We have chosen to deal exclusively with unitary representations since these are the representations that arise in the decomposition of the spaces $L^2(\omega)$. However, it is more standard and in many ways more natural to develop the theory in the category of *admissible* representations. For a discussion of admissible representations, we refer to [Kn], [La] in the archimedean case and to [BZ] and Cartier's article in [Co] in the p-adic case.

Throughout this section, (π, V) will denote an irreducible, unitary representation of G_p on a Hilbert space V. Let $\widehat{G_p}$ be a set of representatives for the equivalence classes of irreducible unitary representations of G_p. The one-dimensional representations of G_p are easily described. They are characters of the form $\chi(\det(g))$, where χ is a unitary character of $\mathbb{Q}_p^*$. All other irreducible unitary representations of G_p are infinite-dimensional (if $\dim(V)$ is finite, the image of π lies in a compact unitary group, but there do not exist any non-trivial finite-dimensional representations of $SL_2(\mathbb{Q}_p)$ with compact image).

The infinite-dimensional representations of G_p are divided into two types:

(1) representations induced from B_p;

(2) square integrable representations.

We recall that π is said to be *square integrable* (or to belong to the *discrete series*) if there exist vectors $v, w \in V_\pi$ such that the matrix coefficient $\langle\pi(g)v, w\rangle$ is square integrable modulo the center:

$$\int_{Z_p\backslash G_p} |\langle\pi(g)v, w\rangle|^2 \, d\dot{g} < \infty.$$

Here $\langle\, , \rangle$ is the inner product on V_π.

Representations induced from B_p

Let

$$\chi : B_p \to \mathbb{C}^*$$

be a continuous character. Since N_p is the commutator subgroup of B_p, χ factors through the quotient $M_p = B_p/N_p$, and thus we may identify χ with a pair of characters (χ_1, χ_2) where χ_j is a character of $\mathbb{Q}_p^*$:

$$\chi\left(\begin{pmatrix} a & b \\ 0 & d \end{pmatrix}\right) = \chi_1(a)\chi_2(d).$$

Let V_χ be the Hilbert space of measurable functions

$$f : G_p \to \mathbb{C}$$

such that

$$f\left(\begin{pmatrix} a & b \\ 0 & d \end{pmatrix} g\right) = \chi_1(a)\chi_2(d) \left|\frac{a}{d}\right|^{\frac{1}{2}} f(g) \tag{2.1}$$

relative to the norm

$$||f||^2 = \int_{K_p} |f(k)|^2 \, dk < \infty. \tag{2.2}$$

Let i_χ denote the action of G on V_χ by right translation. It can be shown that i_χ is unitary if and only if χ itself is a unitary character [La]. The factor $\left|\frac{a}{d}\right|^{\frac{1}{2}}$ in (2.1) is included to make this statement hold. Furthermore, we have the following basic theorem.

Theorem 2.1. *If χ is a unitary character, then (i_χ, V_χ) is an irreducible unitary representation of G.*

When χ is unitary, (i_χ, V_χ) is said to belong to the *unitary principal series*. Certain non-unitary characters χ also give rise to unitary representations. For these χ, i_χ is *unitarizable*, in the sense that there exists a norm on a dense subspace V_χ^0 of V_χ (different from the one defined by (2.2)) preserved by the action of G. The completion of V_χ^0 relative to the new norm is a Hilbert space on which G acts unitarily. Representations of this type are said to belong to the *complementary series*. To describe them, write χ_1 and χ_2 in the form

$$\chi_1(t) = \epsilon_1(t)|t|^{\sigma_1}, \qquad \chi_2(t) = \epsilon_2(t)|t|^{\sigma_2},$$

where ϵ_j are unitary characters of $\mathbb{Q}_p^*$ and $\sigma_j \in \mathbb{R}$.

Theorem 2.2. *Assume that χ is not unitary. Set $\sigma = \sigma_1 - \sigma_2$. Then (i_χ, V_χ) is unitarizable if and only if the following conditions are satisfied:*

(i) $\epsilon_1 = \epsilon_2$

(ii) $\sigma_1 + \sigma_2 = 0$

(iii) $0 < |\sigma| < 1$.

In the case of complementary series, we shall write (i_χ, V_χ) to denote the unitarized representation. The next theorem describes all equivalences among principal and complementary series. If $\chi = (\chi_1, \chi_2)$, let $\chi^w = (\chi_2, \chi_1)$.

Theorem 2.3. *Let χ and χ' be characters of B_p defining unitary principal series or complementary series. Then i_χ is equivalent to $i_{\chi'}$ if and only if $\chi = \chi'$ or $\chi^w = \chi'$.*

Tempered representations

A representation (π, V) is said to be *tempered* if it is *almost* L^2 in the sense that there exist $v, w \in V_\pi$ such that

$$\int_{Z_p\backslash G_p} |\langle \pi(g)v, w\rangle|^{2+\epsilon}\, d\dot{g} < \infty$$

for all $\epsilon > 0$. It is clear that a tempered representation is necessarily infinite-dimensional. The next proposition describes the set of tempered representations of G.

Proposition 2.4. *Let π be an infinite-dimensional, irreducible unitary representation of G. Then π is tempered if and only if π does not belong to the complementary series.*

Discrete series in the archimedean case

Assume $p = \infty$. As above, let (π, V) be an irreducible unitary representation of G_∞. A vector $v \in V$ is said to be K-finite if the set $\{\pi(k)v : k \in K\}$ spans a finite-dimensional subspace. Let V^0 be the space of K-finite vectors. Then V^0 is dense in V. Let $L(G_\infty) = M_2(\mathbb{R})$ be the Lie algebra of G_∞. The vectors $v \in V^0$ are *smooth* in the sense that for any element $X \in L(G_\infty)$, the following derivative exists

$$\pi(X)v = \frac{d}{dt}\pi(e^{tX})v\Big|_{t=0}.$$

The space V_0 is stable under the action of $\pi(X)$. This defines a Lie algebra action $L(G_\infty)$ on V^0 which extends by linearity to an action of the complexification $L(G_\infty)_\mathbb{C} = L(G_\infty) \otimes \mathbb{C}$. For proofs of these and other facts quoted below, see [Kn]. We are interested in the action of the element

$$E^- = \begin{pmatrix} 1 & -i \\ -i & -1 \end{pmatrix}$$

in $L(G_\infty)_\mathbb{C}$.

The action of $L(G_\infty)_\mathbb{C}$ can be made explicit in case V is a space of functions on G_∞ (or a coset space of G_∞) and G_∞ acts by right translation. Let L_{E^-} denote the operator on functions corresponding to E^-. Let $\phi \in V^0$ and consider the restriction of ϕ to G_∞^+. Every element $g \in G_\infty^+$ has a unique decomposition

$$g = \begin{pmatrix} t & 0 \\ 0 & t \end{pmatrix}\begin{pmatrix} 1 & x \\ 0 & 1 \end{pmatrix}\begin{pmatrix} \sqrt{y} & 0 \\ 0 & \sqrt{y}^{-1} \end{pmatrix}\begin{pmatrix} \cos\theta & \sin\theta \\ -\sin\theta & \cos\theta \end{pmatrix}$$

where $y, t > 0$. Relative to these coordinates,

$$L_{E^-} = -2iye^{-2i\theta}\frac{\partial}{\partial \overline{z}} + ie^{-2i\theta}\frac{\partial}{\partial\theta}$$

where $\frac{\partial}{\partial \overline{z}} = \frac{\partial}{\partial x} + i\frac{\partial}{\partial y}$.

Suppose that $\phi(g)$ is a function on G_∞ such that $\phi(gr(\theta)) = e^{ik\theta}\phi(g)$. The restriction of ϕ to G_∞^+ can be written in the form

$$\phi\left(\begin{pmatrix} t & 0 \\ 0 & t \end{pmatrix}\begin{pmatrix} 1 & x \\ 0 & 1 \end{pmatrix}\begin{pmatrix} \sqrt{y} & 0 \\ 0 & \sqrt{y}^{-1} \end{pmatrix} r(\theta)\right) = y^{k/2} f(x+iy) e^{ik\theta},$$

where $f(z)$ is a function on $\mathbb{H}$. By direct calculation, we find

$$L_{E^-}\phi = -2iy^{1+k/2} e^{(k-2)i\theta} \frac{\partial}{\partial \overline{z}} f(x+iy). \tag{2.3}$$

There is a series of square-integrable representations (π_k, V_k) for $k \geq 2$. We call π_k the discrete series representation of weight k. These representations are characterized in the next proposition.

Proposition 2.5. *(1) There is a vector $e_k \in V_k$, unique up to multiples, such that*

(a) $\pi_k\left(\begin{pmatrix} a & b \\ -b & a \end{pmatrix}\right) e_k = \left(\frac{a+bi}{|a+bi|}\right)^k e_k$

(b) $\pi_k(E^-)e_k = 0$.

(2) If (π, V) is an irreducible unitary representation of G_∞ containing a vector $v \in V$ satisfying (a) and (b), then π is unitarily equivalent to π_k.

(3) Every square-integrable representation of G_∞ is of the form $\pi_k \otimes \chi$ for some $k \geq 2$ and some unitary character χ of G_∞.

See [HT], [Kn], or [La] for an explicit construction of π_k and for proof of the facts cited in this proposition.

p-adic case

Now assume $p < \infty$. Square-integrable representations are generally much more difficult to construct in the p-adic case, apart from the special class of *Steinberg representations* (called «special representations» in [JL]). These admit a simple description as quotients of induced representations. For any character ω of $\mathbb{Q}_p^*$, let χ be the character

$$\chi\left(\begin{pmatrix} a & b \\ 0 & d \end{pmatrix}\right) = \omega(ad)\left|\frac{a}{d}\right|^{-\frac{1}{2}}.$$

It is an immediate consequence of the definitions that the function $\omega(\det(g))$ belongs to V_χ and spans a one-dimensional G_p-invariant subspace. Denote the quotient of V_χ by this invariant subspace by $St(\omega)$. Then $St(\omega)$ is irreducible, unitarizable and square-integrable [Go].

Definition. *An irreducible unitary representation (π, V) of G is said to be supercuspidal if there exists a matrix coefficient of π whose support modulo Z is compact.*

The Steinberg representations are not supercuspidal. The next result shows that the main task is to classify supercuspidal representations.

Proposition 2.6. *Let (π, V) be a square-integrable representation of G_p $(p < \infty)$. Then either π is a Steinberg representation, or π is supercuspidal.*

The parametrization of the supercuspidal representations of G_p turns out to be remarkably subtle. Let $\overline{\mathbb{Q}}_p$ be an algebraic closure of $\mathbb{Q}_p$, and let $\Gamma_p = \mathrm{Gal}(\overline{\mathbb{Q}}_p/\mathbb{Q}_p)$ be the absolute Galois group of $\mathbb{Q}_p$. The group Γ_p acts on the residue field $\overline{\mathbb{F}}_p$ of $\overline{\mathbb{Q}}_p$, and this yields a natural map

$$\begin{aligned} \Gamma_p &\longrightarrow \mathrm{Gal}(\overline{\mathbb{F}}_p/\mathbb{F}_p) \\ \sigma &\longrightarrow \overline{\sigma}, \end{aligned}$$

whose kernel is the inertia subgroup $\mathcal{I}$. Let $\Phi \in \mathrm{Gal}(\overline{\mathbb{F}}_p/\mathbb{F}_p)$ be the Frobenius automorphism $\Phi(x) = x^p$. The *Weil group* W_p is the subgroup of elements $\sigma \in \Gamma_p$ such that $\overline{\sigma}$ is an integral power of Φ. The integral powers of Φ are dense in $\mathrm{Gal}(\overline{\mathbb{F}}_p/\mathbb{F}_p)$, and thus W_p is a dense subgroup of Γ_p. However, W_p is given the structure of a locally compact group by declaring that $\mathcal{I}$ and its open subgroups give a neighborhood basis for the identity. Let R_2 be the set of equivalence classes of continuous, irreducible representations

$$\sigma : W_p \longrightarrow GL_2(\mathbb{C}).$$

Let S_2 be the set of (equivalence classes of) supercuspidal representations of G_p. The Langlands correspondence asserts that there is a natural correspondence between

$$R_2 \leftrightarrow S_2.$$

The precise correspondence is stated in terms of the *epsilon factor* $\epsilon(\pi)$ (a certain nonzero complex number) attached to each irreducible unitary representation of G_p by Jacquet and Langlands. It is defined in terms of the Kirillov model for π [JL], [Go]. By a theorem of Langlands, it is also possible to associate an epsilon factor $\epsilon(\sigma)$ to each finite-dimensional, continuous representation of W_p [D2], [Ta]. Local class field theory yields a canonical isomorphism $W_p^{ab} \xrightarrow{\sim} \mathbb{Q}_p^*$ and therefore any character of $\mathbb{Q}_p^*$ may be regarded as a character of W_p. The objects π and σ correspond under the Langlands correspondence if and only if

$$\epsilon(\pi \otimes \chi) = \epsilon(\sigma \otimes \chi)$$

for all characters χ of $\mathbb{Q}_p^*$. Here $\pi \otimes \chi$ denotes the twist of π by the character $\chi(\det(g))$ of G_p.

Various cases of the Langlands correspondence for $GL(2)$ were verified by various authors (cf. [Tu]). The case $p \neq 2$ is already present in [JL]. The most difficult case is when F is a p-adic field of residual characteristic 2. This case was established in full generality by P. Kutzko. A Langlands correspondence is also conjectured for $GL(n)$ over any local field F. We refer to [Ku] for definitions and a description of what is known for general n.

Unramified representations

Assume $p < \infty$. An irreducible unitary representation (π, V) is said to be *unramified* if there exists a nonzero vector $x_p \in V$ fixed under $K_p = GL_2(\mathbb{Z}_p)$.

A character of $\mathbb{Q}_p^*$ is unramified if its kernel contains $\mathbb{Z}_p^*$. We say that a character $\chi = (\chi_1, \chi_2)$ of B_p is unramified if the χ_j are unramified. If χ is unramified, then

$$\chi\left(\begin{pmatrix} a & b \\ 0 & d \end{pmatrix}\right) = \chi_1(p)^{\mathrm{val}(a)}\chi_2(p)^{\mathrm{val}(d)}. \tag{2.4}$$

It follows from the Iwasawa decomposition that i_χ contains a nonzero K_p-fixed vector if and only if χ is unramified. Set

$$f_0(g) = |\frac{a}{d}|^{\frac{1}{2}}\chi_1(p)^{\mathrm{val}(a)}\chi_2(p)^{\mathrm{val}(d)}, \quad \text{where} \quad g = \begin{pmatrix} a & b \\ 0 & d \end{pmatrix} k$$

is an Iwasawa decomposition of g. This function is well-defined if χ is unramified, and in this case, it spans the one-dimensional space of K_p-fixed vectors in V_χ.

Proposition 2.7. *Suppose that π is an infinite-dimensional unramified unitary representation of G. Then there exists an unramified character χ of B such that $\pi \xrightarrow{\sim} i_\chi$. In particular, the space of K_p-fixed vectors is one-dimensional.*

The Hecke operator

Let $C_c^\infty(G_p)$ be the convolution algebra of smooth, compactly-supported functions f on G_p. When $p < \infty$, f is said to be smooth if there exists an open compact subgroup J of G_p such that $f(kgk') = f(g)$ for all $k, k' \in J$. Let (π, V) be an irreducible unitary representation. For $f \in C_c^\infty(G_p)$, we define an operator $\pi(f)$ on V by the operator-valued integral

$$\pi(f) = \int_G f(g)\pi(g)\, dg.$$

This depends on a choice dg of Haar measure, which we regard as fixed, once and for all. It follows immediately from this definition that if f is bi-invariant under an open compact subgroup J, then the image of $\pi(f)$ is contained in the space of J-invariant vectors.

Assume now that $p < \infty$. Let $T_p' \in C_c^\infty(G_p)$ be the characteristic function of the double coset

$$K_p \begin{pmatrix} p & 0 \\ 0 & 1 \end{pmatrix} K_p$$

and set $\widetilde{T}_p = \mathrm{meas}(K_p)^{-1}T_p'$. The operator $\pi(\widetilde{T}_p)$ maps V onto the space of K_p-invariants. It follows that if $\pi = i_\chi$ is unramified, then f_0 must be an eigenvector of $\widetilde{T}_p$ (since the space of K_p-invariants is one-dimensional):

$$i_\chi(\widetilde{T}_p)(f_0) = \mu_p f_0$$

for some $\mu_p \in \mathbb{C}$.

Lemma 2.8. $\mu_p = p^{\frac{1}{2}}(\chi_1(p) + \chi_2(p))$.

Proof. Since $f_0(1) = 1$, we have $\mu_p = i_\chi(\widetilde{T}_p)(f_0)(1)$. By the theory of elementary divisors, there is a decomposition of the double coset as a union of single cosets ([Se])

$$K_p \begin{pmatrix} p & 0 \\ 0 & 1 \end{pmatrix} K_p = \begin{pmatrix} 1 & 0 \\ 0 & p \end{pmatrix} K_p \cup \bigcup_{0 \leq j \leq p-1} \begin{pmatrix} p & j \\ 0 & 1 \end{pmatrix} K_p.$$

Since K_p fixes f_0,

$$\begin{aligned} i_\chi(\widetilde{T}_p)(f_0)(1) &= \operatorname{meas}(K_p)^{-1} \int_{K_p \begin{pmatrix} p & 0 \\ 0 & 1 \end{pmatrix} K_p} f_0(g)\, dg \\ &= f_0(\begin{pmatrix} 1 & 0 \\ 0 & p \end{pmatrix}) + \sum_{0 \leq j \leq p-1} f_0(\begin{pmatrix} p & j \\ 0 & 1 \end{pmatrix}) \\ &= p^{\frac{1}{2}} \chi_2(p) + p^{\frac{1}{2}} \chi_1(p) \end{aligned}$$

as claimed. □

Proposition 2.9. *Let $\pi = i_\chi$ be a representation in either the unitary principal series or the complementary series. Then*

(a) *π is in the unitary principal series if and only if $|\mu_p| \leq 2\sqrt{p}$.*
(b) *π is in the complementary series if and only if $2\sqrt{p} < |\mu_p| < p + 1$.*

Proof. If π is in the complementary series, then $\chi_1(p) = p^{\sigma/2+it}$ and $\chi_2(p) = p^{-\sigma/2+it}$, where $0 < |\sigma| < 1$. Hence

$$2 < |\chi_1(p) + \chi_2(p)| = p^{\sigma/2} + p^{-\sigma/2} < \sqrt{p} + \frac{1}{\sqrt{p}}.$$

On the other hand, if π is in the unitary principal series, then $|\chi_1(p)| = |\chi_2(p)| = 1$ and hence $|\chi_1(p) + \chi_2(p)| \leq 2$. □

Factorization

It is a basic fact that every irreducible unitary representation of $G(\mathbb{A})$ factors as a «restricted» tensor product of representations of the groups G_p. To define restricted tensor products, for all primes p let (π_p, V_p) be a unitary representation of G_p such that π_p is unramified for p outside a finite set S of primes (containing the prime ∞). For $p \notin S$, select a unit vector $x_p \in V_p$ fixed by K_p. Let V_0 be the linear span of the infinite tensors $\otimes v_p$ such that $v_p = x_p$ for almost all p. Then $G(\mathbb{A})$ acts on the tensors $\otimes v_p$ componentwise and hence also on V_0 by linearity. This action is unitary relative to the inner product

$$|| \otimes v_p|| = \prod ||v_p||.$$

Let $\otimes' V_p$ be the Hilbert space completion of V_0. We denote the representation of $G(\mathbb{A})$ on $\otimes' V_p$ by $\otimes \pi_p$.

Theorem 2.10. ([F], [Go], [GGPS]) *Let π be an irreducible unitary representation of $G(\mathbb{A})$. Then there exist irreducible unitary representations π_p for all p, such that π_p is unramified for almost all $p < \infty$, and π is equivalent to $\otimes \pi_p$. Furthermore, the local components π_p are unique up to equivalence.*

According to this theorem, it makes sense to speak of the local components π_p for any irreducible unitary representation of $G(\mathbb{A})$. The following so-called «strong multiplicity one» theorem plays an important role in the theory of automorphic forms.

Theorem 2.11. ([JL]) *Let (π, V) and (π', V') be a cuspidal representation of $G(\mathbb{A})$. Suppose that $\pi_p \xrightarrow{\sim} \pi'_p$ for almost all p. Then $(\pi, V) = (\pi', V')$.*

We have phrased the conclusion of this theorem in terms of equality $(\pi, V) = (\pi', V')$ rather than equivalence of representations. This is because π and π' are not merely equivalent – the subspaces V and V' of $L_0^2(\omega)$ coincide (as they must by Theorem (1.1)).

Modular forms revisited

Let $f \in S_k(N, \omega')$ and let ϕ_f be the associated function on $G(\mathbb{A})$. The next lemma describes the relation between the classical Hecke operator T_p and the operator $\widetilde{T}_p$ for $(p, N) = 1$. Let $\rho = \rho_\omega$ denote the action of $G(\mathbb{A})$ by right-translation on $L^2(\omega)$.

Lemma 2.12. $p^{\frac{k}{2}-1}\rho(\widetilde{T}_p)\phi_f = \phi_{T_p f}$ *for* $(p, N) = 1$.

Proof. By the decomposition in the proof of Lemma 2.8, we have

$$\rho(\widetilde{T}_p)\phi_f(g) = \phi_f(g \begin{pmatrix} 1 & 0 \\ 0 & p \end{pmatrix}_p) + \sum_{0 \le j \le p-1} \phi_f(g \begin{pmatrix} p & j \\ 0 & 1 \end{pmatrix}_p).$$

The subscript p indicates that the adèlic matrix has a component only in the G_p-factor of $G(\mathbb{A})$. We may assume, without loss of generality, that $g = (g_\infty, 1, 1, \ldots)$. Let $\gamma = \begin{pmatrix} 1 & 0 \\ 0 & p^{-1} \end{pmatrix} \in G(\mathbb{Q})$. If $z = g_\infty(i)$, then

$$\phi_f(g \begin{pmatrix} 1 & 0 \\ 0 & p \end{pmatrix}_p) = \phi_f(\gamma g \begin{pmatrix} 1 & 0 \\ 0 & p \end{pmatrix}_p) = \phi_f(\gamma_\infty g_\infty \times \gamma^{p,\infty})$$
$$= \phi_f(\begin{pmatrix} \sqrt{p} & 0 \\ 0 & \sqrt{p}^{-1} \end{pmatrix} g_\infty \times \gamma^{p,\infty})$$

where $\gamma^{p,\infty}$ is the adèlic matrix whose component at ∞ and p is the identity and is γ_ℓ at the remaining primes ℓ. Since $\omega(\gamma^{p,\infty}) = \omega'(p)^{-1}$, we have

$$\phi_f(g \begin{pmatrix} 1 & 0 \\ 0 & p \end{pmatrix}_p) = p^{\frac{k}{2}} \omega'(p)^{-1} j(g_\infty, i)^{-k} f(pz)$$

by (1.5). Similarly $\phi_f(g\begin{pmatrix} p & j \\ 0 & 1 \end{pmatrix}_p) = p^{-\frac{k}{2}} j(g_\infty, i)^{-k} f(\frac{z-j}{p})$. The lemma follows easily. □

We can now make the correspondence between cusp forms and representations. Recall that the Hecke operators T_p for $(p,N) = 1$ act diagonally on $S_k(N,\omega')$ and mutually commute. Hence $S_k(N,\omega')$ has a basis of eigenfunctions for the operators $\{T_p\}_{(p,N)=1}$. For $f \in S_k(N,\omega')$, let $V(f)$ be the closed subspace of $L_0^2(\omega)$ generated by ϕ_f under the action of $G(\mathbb{A})$.

Proposition 2.13. *Let $f \in S_k(N,\omega')$. Suppose that $T_p f = \lambda_p f$ for all p such that $(p,N) = 1$. Then $V(f)$ is an irreducible invariant subspace of $L_0^2(\omega)$. Let π be the representation of $G(\mathbb{A})$ on $V(f)$. Then*

(i) $\pi_\infty = \pi_k$

(ii) *If $(p,N) = 1$, then π_p is unramified and $\pi_p = i_\chi$ where χ is determined by the conditions*

$$\lambda_p = p^{\frac{k-1}{2}}(\chi_1(p) + \chi_2(p))$$
$$\omega_p(p) = \chi_1(p)\chi_2(p).$$

In particular, the set of roots $\{\alpha_p, \beta_p\}$ of (1.10) and the set of numbers $\{p^{\frac{k-1}{2}}\chi_1(p), p^{\frac{k-1}{2}}\chi_2(p)\}$ coincide.

Furthermore, if f' is another cusp form in $S_k(N,\omega')$ such that $T_p f' = \lambda'_p f'$ for $(p,N) = 1$, then $V(f) = V(f')$ if and only if $\lambda_p = \lambda'_p$ for almost all p.

Proof. The representation $V(f)$ is completely reducible since it is a subrepresentation of $L_0^2(\omega)$. Let W be an irreducible subrepresentation of $V(f)$, and let $\tau = \otimes\tau_p$ be the representation of $G(\mathbb{A})$ on W. Let ϕ_W be the orthogonal projection of ϕ_f onto W. Then $L_{E^-}\phi_W = 0$ and

$$\phi_W\left(g\begin{pmatrix} a & b \\ -b & a \end{pmatrix}_\infty\right) = \left(\frac{a+bi}{|a+bi|}\right)^k \phi_W(g)$$

since the projection commutes with the group and Lie algebra action and since these properties hold for ϕ_f (by (2.3) and (1.6)). Thus τ_∞ contains a vector satisfying conditions (a) and (b) of Proposition (2.5) and therefore $\tau_\infty = \pi_k$. Similarly, ϕ_W is invariant under K_p for p not dividing N, and hence τ_p is unramified. We have $\widetilde{T}_p\phi_W = p^{1-\frac{k}{2}}\lambda_p\phi_W$ since this is true for ϕ_f and thus $\tau_p = i_\chi$ as in (ii). Since there is at most one cuspidal representation which satisfies (ii) by Theorem 2.11, W must coincide with V. The last statement of the proposition follows for the same reason. □

Given a cuspidal representation (π, V) such that π_∞ is square-integrable, we can attach a Hecke eigenform to π in the following way. Twisting π by a one-dimensional character if necessary, we may assume that $\pi_\infty = \pi_k$ for some k.

Let ω be the central character of π. Let e_∞ be the vector in V_∞ which satisfies conditions (a) and (b) of Proposition 2.5. Let S be the set of primes such that π_p is ramified (i.e., not unramified) and for $p \notin S$, let e_p be a vector fixed by K_p. Relative to this choice, we have $\pi = \otimes\pi_p$. By a result of W. Casselman (see [D3]), for each $p \in S$, there is an integer $\delta(p) > 0$ called the conductor of π_p, and a so-called «new» vector v_p, unique up to multiples, such that

$$\pi_p\left(\begin{pmatrix} a & b \\ c & d \end{pmatrix}\right) v_p = \omega_p(d) v_p$$

for all $\begin{pmatrix} a & b \\ c & d \end{pmatrix} \in G(\mathbb{Z}_p)$ such that $c \equiv 0 \bmod p^{\delta(p)}$. For $p \in S$, let $e_p = v_p$ and let $\phi \in V$ be the image of the tensor $\otimes e_p$. Then ϕ is a cusp form which, by (2.3), corresponds to a holomorphic cusp form $f \in S_k(N, \omega')$, where $N = \prod p^{\delta(p)}$. The cusp form f is a Hecke eigenform. Suppose $T_p f = \lambda_p f$. By Proposition 2.9 and Proposition 2.13 (ii), the following are equivalent for $(p, N) = 1$:

(1) $|\lambda_p| \leq 2p^{\frac{k-1}{2}}$;
(2) π_p is tempered.

Condition (2) is the representation-theoretic interpretation of the RP conjecture due to Satake [Sa]. We may therefore restate Theorem 1.2 in the following way.

Theorem 2.14. *Let π be a cuspidal representation of $G(\mathbb{A})$ such that π_∞ is square-integrable. Then π_p is tempered for all finite primes p.*

It is conjectured more generally that if π is any cuspidal representation of $GL(2)$, then no local component π_p belongs to the complementary series. However, this conjecture is not known for π such that π_∞ is not square-integrable.

A.3 Quaternion algebras

We recall some standard facts about quaternion algebras (see [V] for details). A quaternion algebra over a field F is a simple algebra D with center F such that $\dim_F(D) = 4$. We say that D is *split* if it is isomorphic to the algebra $M_2(F)$ of 2×2 matrices over F. If $F = \mathbb{Q}_p$ (or more generally, a local field), then up to isomorphism, there are only two quaternions algebras: the split algebra $M_2(F)$ and a unique quaternion division algebra.

If D is a quaternion algebra over $\mathbb{Q}$, then the localization $D_p = D \otimes_{\mathbb{Q}} \mathbb{Q}_p$ is a quaternion algebra over $\mathbb{Q}_p$. We say that D is ramified at p if D_p is a division algebra. Set

$$S_D = \{p \in S(\mathbb{Q}) : D_p \text{ is a division algebra}\}.$$

The following theorem classifies quaternion algebras over $\mathbb{Q}$ (a similar classification holds over any number field).

Theorem 3.1. *(1) Let D be a quaternion algebra over $\mathbb{Q}$. Then S_D is a finite set of even cardinality.*

(2) Let S be a finite set of primes such that Card(S) is even. Then up to isomorphism, there exists a unique quaternion algebra D over $\mathbb{Q}$ such that $S_D = S$.

Quaternion algebras can be constructed explicitly as follows. Let E/F be a separable quadratic extension. Denote the non-trivial Galois automorphism of E/F by a bar. Choose $\xi \in F$ and define $D(E, \xi)$ as follows

$$D(E, \xi) = \{\begin{pmatrix} \alpha & \beta \\ \xi\bar{\beta} & \bar{\alpha} \end{pmatrix} : \alpha, \beta \in E\}.$$

Then $D(E, \xi)$ is a quaternion algebra, and $D(E, \xi)$ is a division algebra if and only if $\xi \notin N_{E/F}(E^*)$. Indeed, there exists a non-zero $x \in D(E, \xi)$ such that $\det(x) = 0$ if and only if $\xi \in N_{E/F}(E^*)$, where $N_{E/F}$ is the norm map. If $F = \mathbb{Q}$, $S_{D(E,\xi)}$ is the set of primes p such that ξ is not a norm from $E \otimes \mathbb{Q}_p$.

The unique quaternion division algebra over $\mathbb{Q}_p$ is isomorphic to $D(E, \xi)$, where E is any quadratic extension $E/\mathbb{Q}_p$ and $\xi \in \mathbb{Q}_p$ is any element which is not a norm from E. For $p = \infty$, the unique quaternion division algebra is the algebra of Hamiltonian quaternions ($D(\mathbb{C}, -1)$ gives an explicit presentation). Note that the set of diagonal elements in $D(E, \xi)$ is isomorphic to E. In particular, every quadratic extension of $\mathbb{Q}_p$ embeds in each quaternion algebra over $\mathbb{Q}_p$.

To construct the global quaternion algebra D in part (2) of Theorem 3.1, let $E/\mathbb{Q}$ be a quadratic extension such that p remains prime in E for all $p \in S$ and choose $\xi \in \mathbb{Q} - N_{E/\mathbb{Q}}(E)$ such that ξ is a local norm from $E \otimes \mathbb{Q}_p$ if and only if $p \in S$. Such elements are well-known to exist. Then D is isomorphic to $D(E, \xi)$.

Theorem 3.2. *Let D be a quaternion algebra over $\mathbb{Q}$ and let $E/\mathbb{Q}$ be a quadratic extension. The following are equivalent:*

(1) *E embeds in D.*

(2) *Each $p \in S_D$ remains prime in E.*

(3) *D is isomorphic to $D(E,\xi)$ for some $\xi \in F^*$.*

For the remainder of this section, we fix a quaternion division algebra D over $\mathbb{Q}$. Let G^* denote D^*, viewed as an algebraic group over $\mathbb{Q}$. Concretely, if F is any field extension of $\mathbb{Q}$, $G^*(F) = (D \otimes_{\mathbb{Q}} F)^*$. Thus $G^*(\mathbb{Q}) = D^*$ and $G_p^* = D_p^*$. As before, $G = GL(2)$. Let Z and Z^* be the centers of G and G^*, respectively. They are both isomorphic to $\mathbb{Q}^*$ and we identify them with each other.

Conjugacy classes

Let $E/\mathbb{Q}$ be a quadratic extension. A $\mathbb{Q}$-linear isomorphism $E \to \mathbb{Q}^2$ defines an action of E^* on $\mathbb{Q}^2$ (by multiplication on E) and hence an embedding

$$E^* \to G(\mathbb{Q}).$$

It is unique up to conjugacy. Furthermore, the images of two elements $x, y \in E^*$ are conjugate if and only if $x = y$ or $x = \sigma_E(y)$ where σ_E is the conjugation on E. An element $\gamma \in G(\mathbb{Q})$ is called *elliptic* if it is the image of an element $x \in E^*$ under such an embedding (γ is scalar if it is in the image of an element of F^*).

In the case of a quaternion division algebra, every element of $G^*(\mathbb{Q})$ is in the image of some embedding $E^* \to G^*(\mathbb{Q})$. However, an embedding $E^* \to G^*(\mathbb{Q})$ for a particular quadratic extension E exists if and only if E embeds in D. Such an embedding exists when the primes in S_D remain prime in E and in this case, the embedding is unique up to conjugacy. As in the case of $GL(2)$, $x, y \in E^*$ have conjugate images if and only if $x = y$ or $x = \sigma_E(y)$.

We say that $\gamma^* \in G^*(\mathbb{Q})$ corresponds to $\gamma \in G(\mathbb{Q})$ if they are images of the same element $x \in E^*$ for some embeddings $E^* \to G^*(\mathbb{Q})$ and $E^* \to G(\mathbb{Q})$. Equivalently, γ^* corresponds to γ if γ is semisimple (i.e., diagonalizable) and the sets of eigenvalues of γ and γ^* coincide (to define the eigenvalues of an element in D, realize D as an algebra of matrices $D(E,\xi)$; the definition is independent of the representation). The correspondence defines an injection

$$\{\text{conjugacy classes in } G^*(\mathbb{Q})\} \to \{\text{elliptic conjugacy classes in } G(\mathbb{Q})\}.$$

This correspondence is defined analogously for the groups G_p^* and G_p. Since every quadratic extension of $\mathbb{Q}_p$ embeds in D_p, we obtain a *bijection* between the set of conjugacy classes in G_p^* and the set of elliptic conjugacy classes in G_p.

Local correspondence
Assume that $p \in S_D$. Then G_p^* is compact modulo its center and the continuous, irreducible representations of G_p^* are all finite-dimensional. The local Jacquet-Langlands correspondence is a bijection between the square-integrable unitary representations of G_p and the irreducible, unitary representations of G_p^*.

To define the correspondence, we must recall the definition of the character of a unitary representation. It is known that the operator

$$\pi(f) = \int_{G_p} f(g)\pi(g)\, dg$$

is of trace class for all $C_c^\infty(G_p)$. By a fundamental theorem of Harish-Chandra, there exists a locally-integrable function $\chi_\pi(g)$ on G_p called the *character* of π, such that

$$Tr(\pi(f)) = \int_{G_p} f(g)\chi_\pi(g)\, dg.$$

An element g is called regular semisimple if it has distinct eigenvalues. The function $\chi_\pi(g)$ is real-analytic (locally constant) on the dense subset of regular semisimple elements in G_p in the case $p = \infty$ (resp., p finite). Hence the value $\chi_\pi(g)$ is well-defined for g regular semisimple. Furthermore, χ_π determines π up to isomorphism. We now state the local Jacquet-Langlands correspondence.

Theorem 3.3. *There exists a bijection $\pi^* \to \pi$ between the set of irreducible unitary representations of G_p^* and the set of square-integrable representations of G_p. It is uniquely characterized by the relation*

$$\chi_{\pi^*}(\gamma^*) = -\chi_\pi(\gamma)$$

for all regular elements $\gamma^* \in G_p^*$ and $\gamma \in G_p$ such that $\gamma^* \to \gamma$.

Examples. There are two cases in which the local Jacquet-Langlands correspondence is simple to describe.

(1) Let $p = \infty$. The value of the character of the square-integrable representation π_k of G_p is well-known [Kn], [La]. An elliptic regular element with eigenvalues $z, \bar{z}$ is conjugate to

$$\gamma = \begin{pmatrix} a & b \\ -b & a \end{pmatrix}$$

where $z = a + bi$. We have

$$\chi_{\pi_k}(\gamma) = -(z\bar{z})^{-\frac{k-2}{2}} \frac{z^{k-1} - \bar{z}^{k-1}}{z - \bar{z}}.$$

On the other hand, G_p^* is the group of Hamiltonian quaternions, which we identify with the group of matrices of the form

$$\begin{pmatrix} \alpha & \beta \\ -\overline{\beta} & \overline{\alpha} \end{pmatrix}$$

with $\alpha, \beta \in \mathbb{C}$. Let ρ'_2 be the «identity» two-dimensional representation of G^*_p. Let ρ'_k denote the $(k-1)$-st symmetric power of ρ'_2. Then ρ'_1 is the trivial representation and in general, ρ'_k is an irreducible, k-dimensional representation. The element

$$\gamma^* = \begin{pmatrix} z & 0 \\ 0 & \bar{z} \end{pmatrix}$$

corresponds to γ defined above. The eigenvalues of γ^* on ρ'_k are $z^i \bar{z}^j$ where $i, j \geq 0$ and $i + j = k - 1$. Therefore

$$\chi_{\rho'_k}(\gamma^*) = \frac{z^k - \bar{z}^k}{z - \bar{z}}.$$

The representations ρ'_k are irreducible but not unitary. Therefore, let ρ_k be the twist $\det^{-\frac{k-1}{2}} \otimes \rho'_k$ for $k > 1$ and let ρ_1 be the trivial representation. It is easy to check that

$$\chi_{\rho_{k-1}}(\gamma^*) = -\chi_{\pi_k}(\gamma).$$

Hence ρ_{k-1} corresponds to π_k under the Jacquet-Langlands correspondence. In particular, the trivial representation ρ_1 corresponds to the weight two discrete series representation π_2.

(2) Suppose that $p < \infty$ is finite and let $\pi = St(\omega)$ where ω is a character of F^*. As described in A.2, there is an exact sequence

$$0 \to \omega \circ \det \to V_\chi \to St(\omega) \to 0.$$

Characters are additive on exact sequences, and hence

$$\chi_\pi = \chi_{i_\chi} - \chi_{\omega \circ \det}.$$

The regular elliptic conjugacy classes do not intersect B_p, and by the formula for an induced character ([La]), χ_{i_χ} vanishes on the regular semisimple conjugacy classes that do not intersect B_p, i.e., on the regular elliptic conjugacy classes. This shows that $\chi_\pi(\gamma) = -\omega(\det(\gamma))$ for γ regular elliptic, and hence that π corresponds to the one-dimensional representation $\omega(\det(g))$ of G^*. Note that π_2 is the analogue of the Steinberg representation for $p = \infty$.

The global correspondence

As above, ω denotes a Hecke character of $\mathbb{I}$, which we also view also as a character of $Z(\mathbb{Q})\backslash Z(\mathbb{A})$. The center Z^* of G^* is isomorphic to Z and thus ω is also viewed as a character of $Z^*(\mathbb{Q})\backslash Z^*(\mathbb{A})$. Let $L^2_*(\omega)$ be the Hilbert space of functions ϕ on $G^*(\mathbb{Q})\backslash G^*(\mathbb{A})$ such that $\phi(zg) = \omega(z)\phi(g)$ and

$$\int_{G^*(\mathbb{Q})Z^*(\mathbb{A})\backslash G^*(\mathbb{A})} |\phi(g)|^2 \, d\dot{g} < \infty,$$

and let ρ^* be the representation of $G^*(\mathbb{A})$ on $L^2_*(\omega)$ by right translation. The quotient space $G^*(\mathbb{Q})Z^*(\mathbb{A})\backslash G^*(\mathbb{A})$ is compact, and this implies that $L^2_*(\omega)$ is completely reducible [GGPS]

$$\rho^* \xrightarrow{\sim} \bigoplus m(\pi^*)\pi^*.$$

Furthermore, the multiplicities $m(\pi^*)$ are finite.

A representation π^* of $G^*(\mathbb{A})$ is called *discrete* if it occurs in the decomposition of ρ^*. Let be the reduced norm on D: $\mathrm{Nm}(d) = \det(i(d))$ where $i : D \to M_2(E)$ is a two-dimensional representation of D. If χ is a Hecke character of $\mathbb{I}$ such that $\chi^2 = \omega$, then the function $\chi \circ \mathrm{Nm}(g)$ spans a one-dimensional invariant subspace of $L^2_*(\omega)$. Every finite-dimensional discrete representation is of this form.

If $p \notin S_D$, then D_p is isomorphic to $M_2(\mathbb{Q}_p)$ and we fix an isomorphism $i_p : G^*(\mathbb{Q}_p) \to G(\mathbb{Q}_p)$ which extends the identification of their centers. This isomorphism is not unique, but any two differ by an inner automorphism. Therefore, we may identify the equivalence classes of irreducible unitary representations of $G^*(\mathbb{Q}_p)$ with those of $G(\mathbb{Q}_p)$. We shall write $\pi_p^* \to \pi_p$ if π_p^* and π_p correspond in this way. Thus the map $\pi_p^* \to \pi_p$ is defined for all primes p. In the statement of the next theorem, we use the fact that unitary representations π^* of $G^*(\mathbb{A})$ are factorizable, with the restricted tensor product defined as in the case of G. The local components are uniquely determined. The next theorem is the global Jacquet-Langlands correspondence.

Theorem 3.4. **([JL])** *Let π^* be a discrete representation of G^*. Assume that π^* is not one-dimensional. Let $\pi = \otimes\pi_p$, where $\pi_p^* \to \pi_p$ for all p. Then π is a cuspidal representation of G. Furthermore, the global map $\pi^* \to \pi$ defines a bijection between the set of infinite-dimensional discrete representations of G^* and the set of cuspidal representations π of G such that π_p is square-integrable for all $p \in S_D$.*

Let π^* be a discrete, infinite-dimensional representation of $G^*(\mathbb{A})$ and suppose that $\pi^* \to \pi$. Then π is cuspidal by Theorem 3.4. If π_∞ is square-integrable, then the RP conjecture holds for π by Deligne's Theorem. Since $\pi_p^* = \pi_p$ for $p \notin S_D$, we obtain the following version of the RP conjecture for G^*. Note that if $p \in S_D$, then G_p^* is compact modulo its center and hence all irreducible unitary representations are tempered.

Theorem 3.5. *Let π^* be a discrete, infinite-dimensional representation of $G^*(\mathbb{A})$. Suppose that $\pi^* \to \pi$ and that π_∞ is square-integrable. Then π_p^* is tempered for all finite primes p.*

Corollary 6.2.2 and Theorem 7.1.1 are special cases of Theorem 3.5.

A.4 The Selberg trace formula

Let ϕ be a smooth function on $G^*(\mathbb{A})$ such that

$$\phi(zg) = \omega^{-1}(z)\phi(g)$$

for all $z \in Z^*(\mathbb{A})$. We assume that ϕ decomposes as a product

$$\phi(g) = \prod_p \phi_p(g_p)$$

where ϕ_p is smooth and compactly-supported modulo Z_p^* for all p, and for almost all p, ϕ_p is the unit in the algebra of bi-K_p-invariant functions on G_p^*, i.e., ϕ_p is the characteristic function of $G^*(\mathbb{Z}_p)$ divided by its measure.

We denote the representation of $G^*(\mathbb{A})$ on $L_*^2(\omega)$ by ρ^*, dropping ω from the notation, and define the operator $\rho^*(\phi)$

$$\rho^*(\phi)f(g) = \int_{Z^*(\mathbb{A})\backslash G^*(\mathbb{A})} \phi(h)f(gh)\,dh.$$

This operator depends on the choice of Haar measure dg on $Z^*(\mathbb{A})\backslash G^*(\mathbb{A})$. It can be represented as an integral operator with kernel:

$$\begin{aligned}\rho^*(\phi)f(g) &= \int_{Z^*(\mathbb{A})\backslash G^*(\mathbb{A})} \phi(h)f(gh)\,dh = \int_{Z^*(\mathbb{A})\backslash G^*(\mathbb{A})} \phi(g^{-1}h)f(h)\,dh \\ &= \int_{Z^*(\mathbb{A})G^*(\mathbb{Q})\backslash G^*(\mathbb{A})} \left(\sum_{\gamma\in Z^*(\mathbb{Q})\backslash G^*(\mathbb{Q})} \phi(g^{-1}\gamma h)\right) f(h)\,dh \\ &= \int_{Z^*(\mathbb{A})G^*(\mathbb{Q})\backslash G^*(\mathbb{A})} K^*(g,h)f(h)\,dh\end{aligned}$$

where

$$K^*(g,h) = \sum_{\gamma\in Z^*(\mathbb{Q})\backslash G^*(\mathbb{Q})} \phi(g^{-1}\gamma h).$$

Since $\rho^*(\phi)$ is an integral operator on a compact space defined by a smooth kernel, it is a trace class operator whose trace is obtained by integration along the diagonal. This integral can be expressed as a sum over a set of representatives $\{\gamma\}$ for the conjugacy classes in $Z^*(\mathbb{Q})\backslash G^*(\mathbb{Q})$, by means of a standard manipulation due to

Selberg:

$$\begin{aligned}\mathrm{Tr}(\rho^*(\phi)) &= \int_{Z^*(\mathbb{A})G^*(\mathbb{Q})\backslash G^*(\mathbb{A})} K^*(g,g)\,dg \\ &= \int_{Z^*(\mathbb{A})G^*(\mathbb{Q})\backslash G^*(\mathbb{A})} \sum_{\gamma\in Z^*(\mathbb{Q})\backslash G^*(\mathbb{Q})} \phi(g^{-1}\gamma g)\,dg \\ &= \sum_{\{\gamma\}} \int_{Z^*(\mathbb{A})G^*_\gamma(\mathbb{Q})\backslash G^*(\mathbb{A})} \phi(g^{-1}\gamma g)\,dg \\ &= \sum_{\{\gamma\}} m_\gamma \int_{G^*_\gamma(\mathbb{A})\backslash G^*(\mathbb{A})} \phi(g^{-1}\gamma g)\,dg.\end{aligned}$$

Here G^*_γ is the centralizer of γ and

$$m_\gamma = \mathrm{meas}(Z^*(\mathbb{A})G^*_\gamma(\mathbb{Q})\backslash G^*_\gamma(\mathbb{A})).$$

For $\gamma \in G^*(\mathbb{A})$, define the orbital integral of ϕ at γ:

$$\Phi(\gamma,\phi) = \int_{G^*_\gamma(\mathbb{A})\backslash G^*(\mathbb{A})} \phi(g^{-1}\gamma g)\,d\dot{g},$$

where $d\dot{g}$ is a quotient measure determined by dg and a choice of measure on $Z^*(\mathbb{A})\backslash G^*_\gamma(\mathbb{A})$. This yields the Selberg trace formula ([GPPS]):

$$\mathrm{Tr}(\rho^*(\phi)) = \sum_{\pi^*} Tr(\pi^*(\phi)) = \sum_{\{\gamma\}} m_\gamma \Phi(\gamma,\phi)$$

An important property of $\Phi(\gamma,\phi)$ is that it is factorizable: if we define the local orbital integral

$$\Phi(\gamma,\phi_v) = \int_{G^*_{\gamma v}\backslash G^*_v} \phi_v(g^{-1}\gamma g)\,d\dot{g},$$

then

$$\Phi(\gamma,\phi) = \prod_v \Phi(\gamma,\phi_v).$$

Selberg Trace formula for $GL(2)$

Now let $G = GL(2)$. Let ρ be the representation of $G(\mathbb{A})$ on the space $L^2(\omega)$. As before, we assume that $\phi = \prod_p \phi_p$ is a factorizable function on $G(\mathbb{A})$ such that

$$\phi(zg) = \omega^{-1}(z)\phi(g)$$

for all $z \in Z(\mathbb{A})$. We again assume that ϕ_p is smooth, compactly-supported modulo Z_p for all p, and that for almost all p, ϕ_p is the characteristic function of $G(\mathbb{Z}_p)$. The operator $\rho(\phi)$ is represented by the integral operator with kernel

$$K(g,h) = \sum_{\gamma \in Z(\mathbb{Q})\backslash G(\mathbb{Q})} \phi(g^{-1}\gamma h).$$

The representation ρ is not completely reducible and in general, $\rho(\phi)$ is not a trace class operator. However, the restriction of $\rho_0(\phi)$ to the space of cusp forms is of trace class. Furthermore, $\rho_0(\phi)$ is represented by a smooth kernel $K_o(g,h)$ and

$$\mathrm{Tr}(\rho_0(\phi)) = \int_{Z(\mathbb{A})G(\mathbb{Q})\backslash G(\mathbb{A})} K_o(g,g)\, dg.$$

The Selberg trace formula provides a useful expression for $Tr(\rho_0(\phi))$, but it is more complicated than in the compact quotient case.

Our exposition (without proofs) of the Selberg trace follows the work of Jacquet-Langlands and Arthur [A], [JL] (cf. also [DL]). Instead of ρ_0, we shall work with the direct sum ρ_d of ρ_0 and the one-dimensional invariant subspaces in $L^2(\omega)$. These latter subspaces are spanned by the functions $\chi(\det(g))$ where χ is a Hecke character of $\mathbb{Q}^*\backslash\mathbb{I}$ such that $\chi^2 = \omega$. Thus,

$$\rho_d = \rho_0 \oplus \left(\bigoplus \chi \circ \det\right).$$

The trace formula has the form

$$Tr(\rho_d(\phi)) = \sum_{\{\gamma\}} J(\gamma,\phi) - \sum_{\{\chi\}} J(\chi,\phi).$$

Here $\{\gamma\}$ is a set of representatives for the *semisimple* (i.e., diagonalizable) conjugacy classes in $Z\backslash G$, and $\{\chi\}$ is a set of representatives for the pairs $\chi = (\chi_1,\chi_2)$ modulo the permutation $(\chi_1,\chi_2) \to (\chi_2,\chi_1)$, where χ_1,χ_2 are Hecke characters of $\mathbb{Q}^*\backslash\mathbb{I}$ trivial on $\mathbb{R}_+^*$ such that $\chi_1\chi_2 = \omega$.

The geometric terms

These are the terms $J(\gamma,\phi)$. There are three types of γ to consider: regular elliptic, regular hyperbolic, and scalar. Recall that a semisimple element is called regular if it has distinct eigenvalues. A regular semisimple element is called elliptic if its eigenvalues generate a quadratic extension of $\mathbb{Q}$ and hyperbolic otherwise. In the hyperbolic case, it is conjugate to a diagonal element in G.

Elliptic regular terms

The elliptic regular terms are the same as in the compact quotient case:

$$J(\gamma,\phi) = m_\gamma \Phi(\gamma,\phi),$$

where $m_\gamma = m(Z(\mathbb{A})G_\gamma(\mathbb{Q})\backslash G_\gamma(\mathbb{A}))$.

Regular hyperbolic terms

In this case, $J(\gamma, \phi)$ is a *weighted* orbital integral. We may assume that $\gamma = \begin{pmatrix} a & 0 \\ 0 & b \end{pmatrix}$ with $a \neq b$. Let $M = G_\gamma$ be the diagonal subgroup. There is a map (not a homomorphism)

$$H : G(\mathbb{A}) \to \mathbb{R}$$

defined as follows in terms of the Iwasawa decomposition. If

$$g = n \begin{pmatrix} \alpha & 0 \\ 0 & \beta \end{pmatrix} k$$

with $n \in N(\mathbb{A})$ and $k \in K$, then

$$H(g) = \frac{1}{2} \log |\frac{\alpha}{\beta}|.$$

It is straightforward to check that $H(g)$ is independent of the choice of Iwasawa decomposition of g. Furthermore, $H(g)$ is a sum of local terms

$$H(g) = \sum_p H_p(g)$$

where $H_p(g)$ is the local function defined in the same way as $H(g)$ using the p-adic absolute value. If $\det(g) = 1$, one has the formulas

$$H_p(\begin{pmatrix} a & b \\ c & d \end{pmatrix}) = \begin{cases} -\frac{1}{2}\ell n(c^2 + d^2) & \text{if } p = \infty \\ -\ell n\,(\max\{|c|, |d|\}) & \text{if } p < \infty. \end{cases}$$

It is immediate from the definition that

$$H(mg) = H(m) + H(g)$$

for all $m \in M(\mathbb{A})$. In particular, the restriction of H to $M(\mathbb{A})$ is a homomorphism. Let

$$M(\mathbb{A})^1 = \{m \in M(\mathbb{A}) : H(m) = 0\}.$$

Define the *weight factor*

$$V(g) = H(g) + H(wg)$$

where $w = \begin{pmatrix} 0 & 1 \\ -1 & 0 \end{pmatrix}$. Then $V(g)$ is a function on $M(\mathbb{A})\backslash G(\mathbb{A})$ since

$$V(mg) = H(mg) + H(wmg) = H(m) + H(g) + H(wmw^{-1}) + H(wg)$$

and $H(wmw^{-1}) = -H(m)$. The weight factor also decomposes as a sum of local terms

$$V(g) = \sum_p V_p(g).$$

Set

$$m_\gamma = m(Z(\mathbb{A})M\backslash M(\mathbb{A})^1).$$

Then

$$J(\gamma, \phi) = m_\gamma \int_{M(\mathbb{A})\backslash G(\mathbb{A})} \phi(g^{-1}\gamma g) V(g)\, dg.$$

This can be written as a sum of terms:

$$J(\gamma, \phi) = m_\gamma \sum_v \left(\prod_{w \neq v} \int_{M_w \backslash G_w} \phi_w(g^{-1}\gamma g)\, dg \right) \int_{M_v \backslash G_v} \phi_v(g^{-1}\gamma g) V_v(g)\, dg.$$

The sum is actually finite because almost all terms vanish. More precisely, if ϕ_p is the unit in the Hecke algebra and $\gamma \in K_p$ (this is the case for almost all p), then $\phi_p(g^{-1}\gamma g)$ vanishes if $g \notin Z_p K_p$ while $V_p(g)$ is zero if $g \in Z_p K_p$.

The term $J(\gamma, \phi)$ for γ scalar is a bit more complicated and so we shall leave them for the end.

The spectral terms

These are the terms $J(\chi, \phi)$ with $\chi = (\chi_1, \chi_2)$. We view χ as a character of $M(\mathbb{A})$:

$$\chi\left(\begin{pmatrix} a & 0 \\ 0 & d \end{pmatrix}\right) = \chi_1(a)\chi_2(d).$$

There are two cases to consider: the regular case $\chi_1 \neq \chi_2$ and the singular case $\chi_1 = \chi_2$.

Regular spectral term

Let $\mathfrak{H}(\chi)$ be the Hilbert space of measurable functions

$$\varphi : N(\mathbb{A})\backslash G(\mathbb{A}) \to \mathbb{C}$$

such that $\varphi(mg) = \chi(m)\varphi(m)$ for $m \in M(\mathbb{A})$ and

$$\int_K |\varphi(k)|^2\, dk < \infty.$$

For $\lambda \in \mathbb{C}$, define an action $I_\lambda = I_{\lambda,\chi}$ of $G(\mathbb{A})$ on $\mathfrak{H}(\chi)$ by the recipe

$$I_\lambda(g)\varphi(x) = e^{-(\lambda+1)H(x)} e^{(\lambda+1)H(xg)} \varphi(xg).$$

This is a realization of the representation induced from the character

$$mn \to \chi(m)e^{\lambda H(m)}$$

of $B(\mathbb{A})$. Thus, the family of representations I_λ are all realized on the same space.

Intertwining operators
Let $w\chi = (\chi_2, \chi_1)$. Define a meromorphic family of intertwining operators

$$M(\lambda) = M(\lambda, \chi) : \mathfrak{H}(\chi) \to \mathfrak{H}(w\chi)$$

for $\lambda \in \mathbb{C}$ as follows. For $Re(\lambda) > 1$, $M(\lambda)$ is defined by an absolutely convergent integral

$$M(\lambda)\varphi(g) = e^{(\lambda-1)H(x)} \int_{N(\mathbb{A})} \varphi(wng)e^{(\lambda+1)H(wng)}\, dn.$$

We have a factorization

$$M(\lambda) = \bigotimes M_v(\lambda)$$

where $M_v(\lambda)$ is the local analogue

$$M_v(\lambda)\varphi(g) = e^{(\lambda-1)H(x)} \int_{N_v} \varphi(wng)e^{(\lambda+1)H(wng)}\, dn,$$

in the sense that if $\varphi = \prod \varphi_v$, then $M(\lambda)\varphi = \prod M_v(\lambda)\varphi_v$. The next theorem is a basic result in the theory of Eisenstein series [L], [GJ].

Theorem 4.1. *The operator $M(\lambda)$ has meromorphic continuation to the entire complex plane. The following hold:*

(i) *$M(\lambda)$ is holomorphic on the unitary line $Re(\lambda) = 0$.*
(ii) *$M(-\lambda, w\chi)M(\lambda, \chi) =$ id.*

The spectral term associated to χ is

$$J(\chi, \phi) = \frac{1}{2\pi} \int_{Re(\lambda)=0} Tr\left(M(-\lambda)M'(\lambda)I_\lambda(\phi)\right) |d\lambda|.$$

Here $M'(\lambda)$ is the derivative of the map $M(\lambda)$:

$$M'(\lambda) = \frac{d}{d\lambda} M(\lambda, \chi) : \mathfrak{H}(\chi) \to \mathfrak{H}(w\chi).$$

The composition $M(-\lambda)M'(\lambda)$ is an endomorphism of $\mathfrak{H}(\chi)$ since

$$M(-\lambda) = M(-\lambda, w\chi) : \mathfrak{H}(w\chi) \to \mathfrak{H}(\chi).$$

It is not an intertwining operator and hence $J(\chi, \phi)$ is not an invariant distribution ($J(\chi, \phi)$ need not equal $J(\chi, \phi^g)$ where $\phi^g(x) = \phi(g^{-1}xg)$).

Normalized intertwining operators

This description of $J(\chi, \phi)$ has the defect that it is not factorizable. The factorization is lost in the process of analytically continuing $M(\lambda)$ (much as the Euler product for the zeta function no longer holds to the left of $\mathrm{Re}(\lambda) = 1$). To recover the product formula, we define *normalized* intertwining operators $R_p(\lambda)$. Fix a continuous additive $\psi : \mathbb{Q}\backslash\mathbb{A} \to \mathbb{C}^*$ and let $\widehat{\chi} = \chi_1\chi_2^{-1}$. Set

$$m_p(\lambda) = \frac{L_p(\lambda, \widehat{\chi})}{L_p(1+\lambda, \widehat{\chi})\epsilon_p(\lambda, \widehat{\chi}, \psi)}$$

where $L_p(\lambda, \widehat{\chi})$ is the local L-factor and $\epsilon_p(\lambda, \widehat{\chi}, \psi)$ is the epsilon factor of Tate's thesis [D2]. Define $R_p(\lambda)$ by the equality

$$M_p(\lambda) = m_p(\lambda)R_p(\lambda).$$

It can be shown that $R_p(\lambda)$ is actually holomorphic in the half-plane $\mathrm{Re}(\lambda) > -1$. The normalization is chosen so that if p is a finite prime such that χ_p is unramified and ψ_p has conductor zero, then

$$R_p(\lambda)\varphi^o_{\chi_p} = \varphi^o_{w\chi_p}$$

where $\varphi^o_{\chi_p}$ is the normalized K_p-fixed vector in $\mathfrak{H}(\chi_p)$ defined by

$$\varphi^o_{\chi_p}(bk) = \chi_p(b).$$

Hence if $\prod_p \varphi_p \in \mathfrak{H}(\chi)$ is a factorizable element ($\varphi_p = \varphi^o_p$ for almost all p), then $\bigotimes R_p(\lambda)\varphi_p$ is a well-defined factorizable element in $\mathfrak{H}(w\chi)$. Furthermore, the derivative satisfies

$$R'_p(\lambda)\varphi^o_{\chi_p} = 0$$

for almost all p.

Globally, set

$$R(\lambda) = \bigotimes R_p(\lambda)$$

$$m(\lambda) = \frac{L(\lambda, \widehat{\chi})}{L(1+\lambda, \widehat{\chi})\epsilon(\lambda, \widehat{\chi})} = \frac{L(1-\lambda, \widehat{\chi}^{-1})}{L(1+\lambda, \widehat{\chi})}.$$

The second inequality follows from the global functional equation of Hecke L-series. In particular, $m(-\lambda) = m(\lambda)^{-1}$. Observe that $\bigotimes R_p(\lambda)$ is well-defined as a product in the region $\mathrm{Re}(\lambda) > -1$. By the additivity of the logarithmic derivative,

$$\begin{aligned} M(-\lambda)M'(\lambda) &= m(-\lambda)R(-\lambda)\frac{d}{d\lambda}(m(\lambda)R(\lambda)) \\ &= m(-\lambda)R(-\lambda)\left(m(\lambda)R'(\lambda) + m'(\lambda)R(\lambda)\right) \\ &= m(-\lambda)m'(\lambda) + R(-\lambda)R'(\lambda). \end{aligned}$$

Furthermore,

$$R(-\lambda)R'(\lambda) = \sum R_p(-\lambda)R'_p(\lambda).$$

The infinite sum is well-defined as an operator because $R'_p(\lambda)\varphi^o_p = 0$ for almost all p as mentioned above. This allows us to write

$$\begin{aligned} J(\chi,\phi) =& \frac{1}{2\pi}\int_{Re(\lambda)=0} \frac{m'(\lambda)}{m(\lambda)} Tr(I_\lambda(\phi))|d\lambda| \\ &+ \sum_p \frac{1}{2\pi}\int_{Re(\lambda)=0} Tr(I_\lambda(\phi^p))Tr(R_p(-\lambda)R'_p(\lambda)I_\lambda(\phi_p))|d\lambda| \end{aligned}$$

where $\phi^p = \prod_{l\neq p}\phi_l$. The sum is again finite. In fact, $Tr(R_p(-\lambda)R'_p(\lambda)I_\lambda(\phi_p)) = 0$ whenever ϕ_p is the characteristic function of K_p divided by its measure (this is the case for almost all p). Indeed, the range of $I_\lambda(\phi_p)$ is then the one-dimensional space spanned by φ^o_p and $R'_p(\lambda)\varphi^o_p$ is independent of λ, so its derivative is zero.

Singular terms

Finally, we describe the terms $J(\gamma,\phi)$ for γ central and $J(\chi,\phi)$ for χ singular.

For $\gamma = \begin{pmatrix} a & 0 \\ 0 & b\end{pmatrix}$, define a distribution locally and globally:

$$\Phi^M(\gamma,\phi) = \left|\frac{a}{b}\right| \int_K\int_N \phi(k^{-1}\gamma n k)\,dn\,dk.$$

If $a \neq b$, then a change of variables shows that

$$\Phi^M(\gamma,\phi) = \left|\frac{(a-b)^2}{ab}\right|^{\frac{1}{2}} \Phi(\gamma,\phi).$$

Let $\zeta_{\mathbb{Q}}(s)$ be the zeta function of $\mathbb{Q}$ and let

$$\zeta_{\mathbb{Q}}(s) = \frac{\lambda_{-1}}{s-1} + \lambda_0 + \cdots$$

be the Laurent expansion of $\zeta_{\mathbb{Q}}(s)$ at $s = 1$.

Now assume $a = b$, so that γ is a scalar. Then for S a sufficiently large set of places of $\mathbb{Q}$ (in particular, for $p \notin S$, ϕ_p is the characteristic function of $G(\mathbb{Z}_p)$ divided by its measure), we have

$$\begin{aligned} J(\gamma,\phi) =& m_\gamma\phi(\gamma) + a_S\Phi^M(\gamma,\phi) \\ &+ \lambda_{-1}\sum_{p\in S}\Phi^M(\gamma,\phi^p)\left(\frac{1}{\zeta_p(1)}\int_{K_p}\int_{\mathbb{Q}_p}\phi(k^{-1}\begin{pmatrix}1 & t\\ 0 & 1\end{pmatrix}k)\ln|t|\,dt\,dk\right) \end{aligned}$$

where

$$a_S = \lambda_0 - \lambda_{-1} \sum_{p \in S} \frac{\zeta_p'(1)}{\zeta_p^2(1)}$$

and $\zeta_p(s)$ is the p-Euler factor of the Riemann zeta function. Thus $\zeta_p(s) = (1 - p^{-s})^{-1}$ for $p < \infty$ and $\zeta_\infty(s) = \pi^{-s/2}\Gamma(s/2)$.

Now let $\chi = (\chi_1, \chi_1)$ be a singular character. Then

$$\begin{aligned} J(\chi, \phi) =& \frac{1}{4\pi} \int_{Re(\lambda)=0} \frac{m'(\lambda)}{m(\lambda)} Tr(I_\lambda(\phi))|d\lambda| \\ &+ \sum_p \frac{1}{4\pi} \int_{Re(\lambda)=0} Tr(I_\lambda(\phi^p))Tr(R_p(-\lambda)R_p'(\lambda)I_\lambda(\phi_p))|d\lambda| \\ &- \frac{1}{4} Tr(M(0)I_0(\phi)). \end{aligned}$$

Observe that since $\chi = w\chi$, $M(\lambda)$ maps $\mathfrak{H}(\chi)$ to itself and hence $Tr(M(0)I_0(\phi))$ is defined.

On the proof of the Jacquet-Langlands correspondence

The proof uses the technique of «comparing trace formulas» introduced in [L] (cf [GJ]). One shows that it is possible to define a correspondence of functions

$$\phi^* \to \phi$$

with the properties

(1) $m_{\gamma^*}\Phi(\gamma^*, \phi^*) = m_\gamma \Phi(\gamma, \phi)$ if γ^* is a regular element in $G^*(\mathbb{A})$ and $\gamma^* \to \gamma$.
(2) $\Phi(\gamma, \phi) = 0$ if γ is a regular semisimple element which does not correspond to any $\gamma^* \in G^*(\mathbb{A})$.

The correspondence $\phi^* \to \phi$ is defined componentwise via a similar locally-defined correspondence $\phi_p^* \to \phi_p$. For $p \notin S_D$, we take $\phi_p = \phi_p^*$. For $p \in S_D$, ϕ_p is a function which satisfies

$$\Phi(\gamma, \phi_p) = \begin{cases} \Phi(\gamma^*, \phi_p^*) & \text{if } \gamma^* \text{ is regular and } \gamma^* \to \gamma \\ 0 & \text{if } \gamma \text{ is a regular hyperbolic element.} \end{cases}$$

The correspondence depends on how the measures used to define the orbital integrals on G_p and G_p^* are normalized. To fix the normalization, it is best to use matching Tamagawa measures on G_p and G_p^* as defined in [JL] (cf. [GJ], p. 246). For γ^* regular, we may identify $G_{\gamma p}$ with $G_{\gamma^* p}^*$ and choose the Haar measures on the two groups to be the same under the identification. Then we obtain matching quotient measures on $G_{\gamma p}\backslash G_p$ and $G_{\gamma^* p}^*\backslash G_p^*$. With these choices, it can be shown that for scalar elements $\gamma^* \in G_p^*$,

$$\phi_p^*(\gamma^*) = -\phi_p(\gamma)$$

if $\gamma^* \to \gamma$. Since S_D has even cardinality, we obtain globally

$$\phi^*(\gamma^*) = \phi(\gamma)$$

for $\gamma \in G(\mathbb{A})$ scalar. Globally, the Tamagawa measures satisfy

$$m_{\gamma^*} = m_\gamma.$$

Since the set S_D is non-empty and of even cardinality, there are at least two primes p such that $\Phi(\gamma, \phi_p) = 0$ for all regular hyperbolic elements. This forces the vanishing of many terms in the trace formula for G. For example, it is immediate that $J(\gamma, \phi) = 0$ for all regular hyperbolic elements. It is not difficult to show that the singular term simplifies to $J(\gamma, \phi) = m_\gamma \phi(\gamma)$ for γ scalar. Furthermore,

$$J(\chi, \phi) = 0$$

for all χ. This follows because $\mathrm{Tr}\,(\pi(\phi_p)) = 0$ for all $p \in S_D$ and all principal series representations π. Indeed, $\mathrm{Tr}(\pi(\phi_p))$ can be expressed in terms of the regular hyperbolic orbital integrals [La]. We conclude that if ϕ corresponds to ϕ^*, then

$$\mathrm{Tr}\,(\rho_d(\phi)) = \sum_{\{\gamma\}} m_\gamma \Phi(\gamma, \phi).$$

Observe also that if γ is not a scalar, then $\Phi(\gamma, \phi) = 0$ unless γ is elliptic regular in G_p for all $p \in S_D$. Thus $\Phi(\gamma, \phi) = 0$ unless γ lies in the image of an embedding of E^* in G, where E is a quadratic extension in which no prime in S_D splits. The latter condition is necessary and sufficient for the existence of an embedding of E^* into $G^*(\mathbb{Q})$, and thus γ corresponds to some $\gamma^* \in G^*(\mathbb{Q})$. It now follows that

$$\begin{aligned}
\mathrm{Tr}\,(\rho_d(\phi)) &= \sum_{\{\gamma\}} m_\gamma \Phi(\gamma, \phi) \\
&= \sum_{\{\gamma^*\}} m_{\gamma^*} \Phi(\gamma^*, \phi) \\
&= \mathrm{Tr}(\rho^*(\phi^*)).
\end{aligned}$$

This completes our sketch of the proof of the following theorem.

Theorem 4.2. *If $\phi^* \to \phi$, then*

$$\sum_\pi \mathrm{Tr}(\pi(\phi)) = \sum_{\pi^*} \mathrm{Tr}\,(\pi^*(\phi^*))$$

where π (resp. π^*) ranges over the discrete representations of $G(\mathbb{A})$ (resp. $G^*(\mathbb{A})$).

The Jacquet-Langlands correspondence can be deduced from this theorem using techniques introduced in [L]; see [GJ]. A slightly different route was chosen in the original proof [JL].

References to the Appendix

[AA] Automorphic forms, Shimura varieties, and L-functions, vols I and II, Edited by: L. Clozel and J. Milne, Academic Press, 1990

[A] J. Arthur, *The Selberg trace formula for groups of F-rank one*, Ann. of Math. **100** (1974), 326–385.

[Bo] *Algebraic groups and discontinuous subgroups*, Proceeding of symposia in pure mathematics IX (Eds. A. Borel and G.D. Mostow), AMS, 1966.

[BR] D. Blasius and J. Rogawski, *Zeta-functions of Shimura varieties*, Conference on Motives, Proceeding of symposia in pure mathematics (Eds. U. Janssen and J.-P. Serre), AMS, 1994.

[BZ] J. Bernstein and A. Zelevinsky, *Representations of the group $GL(n,F)$, where F is a local non-archimedean field.* Russian Math. Surveys **31 (3)** (1976), 5–70.

[CF] J.W.S. Cassels and A. Fröhlich, *Algebraic number theory*, Academic Press, 1967.

[Co] *Automorphic forms, representations, and L-functions*, Proc. Sympos. Pure Math. 33, Parts I and II (Eds. A. Borel and W. Casselman), AMS, Providence, R.I. 1979.

[D1] P. Deligne, *Formes modulaires et représentations ℓ-adique*, Sém. Bourbaki 1969, Exp. 55., SLN 179 Springer-Verlag, 1969, 139–172.

[D2] P. Deligne, *Les constantes des équations fonctionnelles des fonctions L*, Modular functions of one variable. II, SLN 249, Springer-Verlag 1973, 501–595.

[D3] P. Deligne, *Formes modulaires et représentations de $GL(2)$*, Modular functions of one variable. II, SLN 249, Springer-Verlag, 1973.

[DL] M. Duflo and J.-P. Labesse, *Sur la formule des traces de Selberg*, Ann. Sci. de l'Ec. Norm. Sup. **4** (1971), 193–284.

[E] M. Eichler, *Quaternäre quadratische Formen und die Riemannsche Vermutung für die Kongruenzzetafunction*, Archiv d. Math. V (1954), 355–366.

[F] D. Flath, *Decomposition of representations into tensor products*, [Co], Part 1, 179–183.

[G1] S. Gelbart, *Automorphic forms on adele groups*, Annals of Math. Studies, Princeton U. Press Princeton, 1975.

[GJ] S. Gelbart and H. Jacquet, *Forms of $GL(2)$ from the analytic viewpoint*, [Co], Part 1 213–251.

[Go] R. Godement, *Notes on Jacquet-Langlands theory*, Institute for Advanced Study, 1970.

[GGPS] I. Gelfand, M. Graev, and I. I. Piatetski-Shapiro, *Representation theory and automorphic functions*, W.B. Saunders Co., 1969.

[HT] R. Howe and E. Tan, *Non-abelian harmonic analysis*, Universitext, Springer-Verlag, 1992.

[JL] H. Jacquet and R. Langlands, *Automorphic forms on* $GL(2)$, SLN 114, Springer-Verlag, 1970.

[Ka] N. Katz, *An overview of Deligne's proof of the Riemann hypothesis for varieties over finite fields*, Mathematical developments arising from Hilbert problems, Proc. Symp. in Pure Math. XXVIII, AMS, 1976, 275–305.

[Kn] A. Knapp, *Representation theory of semisimple groups by examples*, Princeton U. Press, 1986.

[Kn2] A. Knapp, *Arithmetic of Elliptic curves*, Princeton U. Press, 1992.

[Ku] S. Kudla AMS Conference on Motives, Proceeding of symposia in pure mathematics, AMS, 1994.

[La] S. Lang, $SL_2(\mathbf{R})$, Addison-Wesley, Reading, Mass., 1975.

[L] R.P. Langlands, *Eisenstein series*, SLN 544, Springer-Verlag, 1976.

[Sa] I. Satake, *Spherical functions and Ramanujan conjecture*, Algebraic groups and discontinuous subgroups, Proc. Symp. in Pure Math. IX, AMS, 1966, 258–264.

[Se] J.-P. Serre, *A Course on arithmetic*, Springer-Verlag, 1973.

[Sh1] G. Shimura, *Introduction to the arithmetic theory of automorphic forms*, Princeton U. Press, 1971.

[S2] G. Shimura, *Correspondances modulaires et les fonctions* ζ *des courbes algébriques*, J. Math. Soc. Japan **10** (1958), 1–18.

[Ta] J. Tate, *Number theoretic background*, [Co], Part 2, 3–26.

[Tu] J. Tunnell, *Report on the local Langlands conjecture for* $GL(2)$, [Co], Part 2, 135-138.

[V] M.-F. Vigneras, *Arithmétique des algbres de quaternions*, SLN 800, Springer-Verlag, 1980.

References

[Ab] H. Abelson, *A note on time-space tradeoffs for computing continuous functions*, Inform. Process. Lett. **8** (1979), 215–217.

[AS] S.R. Adams and R.J. Spatzier, *Kazhdan groups, cocycles and trees*, Amer. J. of Math. **112** (1990), 271–287.

[AKS1] M. Ajtai, J. Komlos and E. Szemeredi, *An* $0(n \log n)$ *sorting network*, in Proc. 15th Annual ACM Symposium on the Theory of Computing, Association for Computing Machinery, New York (1983).

[AKS2] M. Ajtai, J. Komlos and E. Szemeredi, *Sorting in* $c \log n$ *parallel steps*, Combinatorica **3** (1983), 1–9.

[A1] N. Alon, *Eigenvalues and expanders*, Combinatorica **6** (1986), 83–96.

[A2] N. Alon, *Eigenvalues, geometric expanders, sorting in rounds and Ramsey theory*, Combinatorica **6** (1986), 207–219.

[A3] N. Alon, *The number of spanning trees in regular graphs*, in: Random Structures and Algorithms **1** (1990), 175–181.

[AGM] N. Alon, Z. Galil and V.D. Milman, *Better expanders and superconcentrators*, J. of Algorithms **8** (1987), 337–347.

[AM] N. Alon, and V.D. Milman, λ_1, *isoperimetric inequalities for graphs and superconcentrators*, J. Comb. Th. **B 38** (1985), 78–88.

[AR] N. Alon, Y. Roichman, *Random Cayley Graphs and Expanders*, Random Structures and Algorithms, to appear (abstract in [Fr3] pp. 1–3).

[Alp] R. Alperin, *Locally compact groups acting on trees and property T*, Mh. Math. **93** (1982), 261–265.

[ACPTTV] J. Angel, N. Celniker, S. Poulos, A. Terras, C. Trimble and E. Velasquez, *Special functions on finite upper half planes*, Contemporary Math. **138** (1992), 1–26.

[An] D. Angluin, *A note on a construction of Margulis*, Inform. Process Letters **8** (1979), 17–19.

[AG] D. Angluin, A. Gardiner, *Finite common covering of pairs of regular graphs*, J. of Comb. Th. B **30** (1981), 184–187.

[AB] F. Annexstein, M. Baumslag, *On the diameter and bisector size of Cayley graphs*, preprint.

[B] L. Babai, *Spectra of Cayley graphs*, J. of Comb. Th. B **27** (1979), 180–189.

[BaHe] L. Babai, G. Hetyei, *On the diameter of random Cayley graphs of the symmetric group*, Combinatorics, Probability and Computing **1** (1992), 201–208.

[BHKLS] L. Babai, G. Hetyei, W.M. Kantor, A. Lubotzky, A. Seress, *On the diameter of finite groups*, 31 IEEE Symp. on Foundations of Computer Science (FOCS 1990), 857–865.

[BKL] L. Babai, W.M. Kantor and A. Lubotzky, *Small diameter Cayley graphs for finite simple groups*, Europ. J. of Combinatorics, **10** (1989), 507–522.

[BDH] R. Bacher, P. de la Harpe, *Exact value of Kazdan constants for some finite groups*, J. of Algebra **163** (1994), 495–515. to appear.

[Ba] S. Banach, *Sur le probleme de la measure*, Fund. Math. **4** (1923), 7–33. Reprinted in: S. Banach, «Oeuvres», Vol. I, Warsaw, Editions Scientifique de Pologne 1967.

[Bss1] H. Bass, *The degree of polynomial growth of finitely generated nilpotent groups*, Proc. Lond. Math. Soc. **25** (1972), 603–614.

[Bss2] H. Bass, *Covering theory for graphs of groups*, J. Pure Appl. Algebra **89** (1993), 3–47.

[BK] H. Bass, R. Kulkarni, *Uniform tree lattices*, J. of A.M.S. **3** (1990), 843–902.

[BL] H. Bass, A. Lubotzky, *Non-uniform tree lattices*, in preperation.

[BMS] H. Bass, J. Milnor, J.P. Serre, *Solution of the congruence subgroup problem for SL_n $(n \geq 3)$ and Sp_{2n} $(n \geq 2)$*, Publ. Math. IHES, **33** (1967), 59–137.

[Bs] L. Bassalygo, *Asymptotically optimal switching circuits*, Problems of Inform. Trans. **17** (1981), 206–211.

[BP] L. Bassalygo and M. Pinsker, *Complexity of an optimum nonblocking switching network without reconnections*, Problemy Peredachi Informatsii, **9** (1) (1973), 84–87; Problems Inform. Transmission, **9** (1) (1974), 64–66.

[Be] A.F. Beardon, «The Geometry of Discrete Groups», Graduate Texts in Mathematics **91**, Springer-Verlag, New York, 1983.

[BGG] M. Bellare, O. Goldreich, S. Goldwasser, *Randomness in interactive proofs*, In Proc. Foun. of Comp. Sci. (FOCS), (1990).

[Bér] P.H. Bérard, «Spectral Geometry: Direct and Inverse Problems», Springer Lecture Notes in Math. **1207**, Springer Verlag, New York 1986.

[BGM] M. Berger, P. Gauduchon and E. Mazet, «*Le Spectré d'une Variété Riemannienne*», Springer Lecture Notes in Math. **194**, Springer-Verlag 1971.

[Beg] M. Berger, *Les placements de cercles*, in: Pour la Science **176**, Juin 1992, pp. 72–79.

[Bi] F. Bien, *Constructions of telephone networks by group representations*, Notices A.M.S. **36** (1) (1989) 5–22.

[Bg] N. Biggs, «Algebraic Graph Theory», Cambridge University Press, London, 1974.

[BB] N.L. Biggs, A.G. Boshier, *Note on the girth of Ramanujan graphs*, J. Combin. Theory B **49** (1990), 190–194

[BH] N. Biggs, M.J. Hoare, *The sextet construction for cubic graphs*, Combinatorica **3** (1983), 153–165.

[BKVPY] M. Blum, R. Karp, O. Vornberger, C. Papadimitriou and M. Yannukakis, *The complexity of testing whether a graph is a superconcentrator*, Inform. Process Letters **13** (1981), 164–167.

[B1] B. Bollobás, «Graph Theory», Springer-Verlag, New York, 1979.

[B2] B. Bollobás, «Extremal Graph Theory», Academic Press, London, 1978.

[B3] B. Bollobás, «Extremal Graph Theory with Emphasis on Probabilistic Methods», CBMS no. 62, Amer. Math. Soc., Providence, R.I., 1986.

[B4] B. Bollobás, «Random Graphs», Academic Press, London 1985.

[B5] B. Bollobás, *The isoperimetric number of random regular graphs*, Europ. J. Combinatorics **9** (1988), 241–244.

[Bo1] A. Borel, *Some finiteness properties of adele groups over number fields*, Publ. Math. IHES **16** (1963), 101–126.

[Bo2] A. Borel, *On free subgroups of semi-simple groups*, Ens. Math. 29 (1983), 151–164.

[BHC] A. Borel, Harish-Chandra, *Arithmetic subgroups of algebraic groups*, Ann. of Math. **75** (1962), 485–535.

[BSh] Z.I. Borevich, I.R. Shafarevich, «Number Theory», Academic Press, New York, 1966.

[BS] A. Broder and E. Shamir, *On the second eigenvalue of random regular graphs*, 28th Annual Symp. on Found. of Comp. Sci. (1987) 286–294.

[Br1] R. Brooks, *Combinatorial problems in spectral geometry*, in: «Curvature and Topology of Riemanian Manifolds», (Eds: K. Shishana, T. Sakai and T. Sunada) Springer Lect. Notes in Math. **1201** (1986), 14–32.

[Br2] R. Brooks, *The spectral geometry of a tower of coverings*, J. of Diff. Geom. **23** (1986), 97–107.

[Br3] R. Brooks, «*Spectral Geometry*», Cambridge University Press, to appear.

[Br4] R. Brooks, *The spectral geometry of k-regular graphs*, J. d'Analyse **57** (1991), 120–151.

[Br5] R. Brooks, *Some relations between spectral geometry and number theory*, in: Topology 90 (ed: Apanosov et.al) 61–75, Walter de Gruyter 1992.

[Br6] R. Brooks, *Trace formula methods in spectral geometry*, in: «Colloque en l'honneur de J.P. Kahane», Astérisque, to appear.

[Br] K.S. Brown, «Buildings», Springer-Verlag, New York, 1989.

[BT] F. Bruhat and J. Tits, *Groupes algebraiques simples sur un corps local*, Proc. Conf. Local Fields, Springer-Verlag, (1967), pp. 23–36.

[Bc] M.W. Buck, *Expanders and diffusers*, SIAM J. Alg. Disc. Math. **7** (1986), 282–304.

[Bur1] M. Burger, *Estimation de petites valeurs propres du Laplacian d'un revetement de varietés Riemanniennes compartes*, C.R.A.S. Paris, **t. 302** Series I, no. 5 (1986), 191–194.

[Bur2] M. Burger, *Grandes valeurs propres du Laplacian et gruphes*, Sem. de The. Spectral et Geometry, Chambiery – Grenoble, 1985/6, 95–100.

[Bur3] M. Burger, *Spectre du Laplacien, graphes et topologie de Fell*, Comm. Math. Helv. **63** (1988), 226–252.

[Bur4] M. Burger, *Cheng's inequality for graphs*, unpublished preprint.

[Bur5] M. Burger, *Kazhdan constants for* $SL_3(Z)$, J. Reine-Angew. Math., **413** (1991), 36–67.

[BuS] M. Burger, P. Sarnak, *Ramanujan duals II*, Invest. Math. **106** (1991), 1–11.

[Bu1] P. Buser, *On Cheeger's inequality* $\lambda_1 \geq \frac{h^2}{4}$, Proc. Symp. Pure Math. **36** (1980), 29–77.

[Bu2] P. Buser, *A note on the isoperimetric constant*, Ann. Sci. Ecole Norm. Sup. **15** (1982), 213–230.

[Bu3] P. Buser, *On the bipartition of graphs*, Discrete Applied Mathematics **9** (1984), 105–109.

[Bu4] P. Buser, *Cubic graphs and the first eigenvalue of a Riemann surface*, Math. Z. **162** (1978), 87–99.

[Ca] D. Cantor, *On nonblocking switching networks*, Networks **1** (1971), 367–377.

[Car] R.W. Carter, «Simple Groups of Lie Type», Wiley-Interscience, London, 1972.

[CMS] D.I. Cartwright, W. Mlotkowski and T. Steger, *Property* (T) *and* $\tilde{A}_2$ *groups*, preprint.

[CPTTV] N. Celniker, S. Poulos, A. Terras, C. Trimble and E. Velasquez, *Is there life on finite upper half planes*, Contemporary Math. **143** (1993), 65–88.

[Ch] I. Chavel, «Eigenvalues in Riemanian Geometry», Pure and Appl. Math. Vol. 115, Academic Press, New York (1984).

[Cr] J. Cheeger, *A lower bound for the smallest eigenvalue of the Laplacian*, in: «Problems in Analysis», Ganning (ed.) Princeton Univ. Press (1970), 195–199.

[Che] S.Y. Cheng, *Eigenvalues Comprasion Theorems and its Geometric Applications*, Math. Z. **143** (1975), 289–297.

[Chi] P. Chiu, *Cubic Ramanujan graphs*, Combinatorica, **12** (1992), 275–285.

[Cho] C. Chou, *Ergodic group actions with non-unique invariant means*, Proc. A.M.S. **100** (1987), 647–650.

[CLR] C. Chou, A.T. Lau and J. Rosenblatt, *Approximation of compact operators by sums of translations*, Ill. J. Math. **29** (1985), 340–350.

[C1] F.R.K. Chung, *On concentrators, superconcentrators, and nonblocking networks*, Bell System Tech. J., **58** (1979), 1765–1777.

[C2] F.R.K. Chung, *Diameters and eigenvalues*, J. of the Amer. Math. Soc., **2** (1989), 187–196.

[C3] F.R.K. Chung, *The Laplacian of a hypergraph*, in [Fr3] pp. 21–36.

[CLPS] J. Cogdell, J. Li, I. Piatetski-Shapiro and P. Sarnak, *Poincaré series for* $SO(n,1)$, Acta-Math. **167** (1991), 229–285.

[CV1] Y. Colin de Verdiere, *Spectres de varietes riemanniennee et spectree de graphes*, Proc. Int. Cong. Math. Berkeley (1986), 522–530.

[CV2] Y. Colin des Verdiere, *Comment distributer de points uniforment sur une sphere (d'apres Lubotzky, Phillips et Sarnak)* Collection, Seminaire de Theorie Spectrale et Geometrie, **5**, (1986–7), pp. 9–18.

[CV3] Y. Colin de Verdiere, *Distribution de points sur une sphere (d'apres Lubotzky, Phillips et Sarnak)*, Seminaire N. Bourbaki, vol. 1988–89, November 1988, Exp: **703**, Asterisque **177–178** (1989), 83–93.

[CW] A. Connes and B. Weiss, *Property* (T) *and asymptotically invariant sequences*, Israel J. Math. **37** (1980), 209–210.

[CS] J.H. Conway, N.J.A. Sloane, «Sphere Packings, Lattices and Groups», Springer Verlag, New York, 1988.

[CDS] D. Cvetkovic, M. Doob and H. Sachs, «Spectra of Graphs: Theory and Applications», Academic Press, New York, 1980.

[DaSa] G. Davidoff, P. Sarnak, *An Elementary Approach to Ramanujan Graphs*, to appear.

[DKi] C. Delaroche and A. Kirillov, *Sur les relations entre l'espace d'un groupe et la structure de ses sous-groupes fermes*, Sem. Bourbaki, Expose 343, 1968. Springer Lect. Notes Math. **180** (1972).

[Dl1] P. Deligne, *Formes modulaires et representations l-adiques*. Sem. Bourbaki 1968/1969 No. 355, in Lecture Notes in Math. **179**, Springer Verlag (1971), 139–172.

[Dl2] P. Deligne, *La conjecture de Weil I*, Publ. Math. I.H.E.S. **43** (1974), 273–308.

[DS] P. Deligne and D. Sullivan, *Division algebras and the Hausdorff-Banach-Tarski paradox*, L'Enseignement Math. **29** (1983), 145–150.

[DJR] A. Del-Junco and J. Rosenblatt, *Counterexamples in ergodic theory and number theory*, Math. Ann. **245** (1979), 185–197.

[D] P. Diaconis, «Group Representation in Probability and Statistics» Inst. of Math. Statistics, Howard, California, 1988.

[DG] P. Diaconis, C. Greene, *Applications of Murphy's elements*, preprint.

[DSh] P. Diaconis, M. Shahshahani, *Generating a random permutation with random transpositions*, Z. Wahrscheinlichkeitstheorie verw. Gebriete **57** (1981), 159–179.

[Di] L.E. Dickson, *Arithmetic of quaternions*, Proc. Lond. Math. Soc. **20** (1922), 225–232.

[Dix] J. Dixon, *The probability of generating the symmetric group*, Math. Z. **110** (1969) 199–205.

[DDMS] J.D. Dixon, M.P.F. de Sautoy, A. Mann D. Segal «Analitic Pro-p Groups», London Math. Soc. LNS **157** Cambridge University Press 1991.

[Do] J. Dodziuk, *Difference equations, Isoperimetric inequality and transcience of certain random walks*, Trans. A.M.S. **284** (1984), 787–794.

[DK] J. Dodziuk and W.S. Kendall, *Combinatorial Laplacians and isoperimetric inequality*, in: «From Local times to global geometry, control and physics» (Ed: K.D. Ellworthy), Pitman Research Notes in Mathematics Series 150, pp. 68–74.

[Dor] L. Dornhoff, «Group Representation Theory», part A, Marcel Dekker Inc., New York, 1971.

[Dr1] V.G. Drinfeld, *Finitely additive measures on S^2 and S^3, invariant with respect to rotations*, Func. Anal. and its Appl. **18** (1984), 245–246.

[Dr2] V.G. Drinfeld, *The proof of Peterson's conjecture for $GL(2)$ over a global field of characteristic p*, Functional Analysis and its applications **22** (1988), 28–43.

[Ei] M. Eichler, *Quaternare quadratische Formen und die Riemannsche Vermutung für die Kongruenz Zetafunktion*, Arch. der Math. 5 (1954), 355–366.

[Ef] I. Efrat, *Automorphic spectra on the tree of PGL_2*, L'Enseignement Mathematique **37** (1991), 31–43.

[ELM] J. Elstrodt, F. Grunewald and J. Mennicke, *Poincare series, Kloosterman sums and eigenvalues of the Laplacians for congruence groups acting on hyperbolic spaces*, C. R. Acad. Sci. Paris **305** (1987), 577–580.

[Ep] D.B.A. Epstein, *Finite presentations of groups and 3-manifolds*, Quart. J. Math. Oxford **12** (1961) 205–212.

[EGS] P. Erdos, R.L. Graham and E. Szemeredi, *On sparse graphs with dense long paths*, Comput. Math. Appl. **1** (1975), 365–369.

[ES] P. Erdös and H. Sachs, *Reguläre Graphen gegebener Taillenweite mit minimaler Knollenzahl*, Wiss. Z. Univ. Halle-Willenberg Math. Nat. R. **12** (1963), 251–258.

[FH] W. Feit and G. Higman, *The nonexistence of certain generalized polygons*, J. of Alg. **1** (1964), 114–131.

[FTN] A. Figa-Talamanka, C. Nebia, «Harmonic Analysis and Representation Theory for Groups Acting on Homogenous Trees», London Math. Soc. LNS **162**, Cambridge University Press (1991).

[FTP] A. Figa-Talamanca, M.A. Picardello, «Harmonic Analysis on Free Groups», Lect. Notes Pure Appl. Math. **87**, Marcel Dekker, New York 1983.

[FFP] P. Feldman, J. Friedman and N. Pippenger, *Wide sense nonblocking networks*, SIAM J. Disc. Math. **1** (1988), 158–173.

[Fe] J.M.G. Fell, *Weak containment and induced representations of groups*, Canadian J. Math. **14** (1962), 237–268.

[FOW] L. Flatto, A. Odlysko and D. Wales, *Random shuffles and group representations*, Ann. of Probability **13** (1985), 154–178.

[Fr1] J. Friedman, *On the second eigenvalue and random walk in random d-regular graph*, Combinatorica **11** (1991), 331–362.

[Fr2] J.Friedman, *The spectra of infinite hypertrees*, SIAM J. Comp. **20** (1991) 951–961.

[Fr3] J. Friedman, *Expanding graphs*, Proceeding of a DIMACS Workshop May 11–14, 1992. DIMACS Series in Diserte Mathematics and Theoretical Computer Science, Vol. 10, American Math. Soc. 1993.

[Fr4] J. Friedman, *Some geometric aspects of graphs and their eigenfunction*, Duke Math. J. **69** (1993), 487–525.

[FKS] J. Friedman, J. Kahn and E. Szemerédi, *On the second eigenvalue in random regular graphs*, Proc. 21st ACM STOC, ACM press (1989), 587–598.

[FP] J. Friedman and N. Pippenger, *Expanding graphs contain all small trees*, Combinatorica **7** (1987), 71–76.

[FW] J. Friedman, A. Wigderson, *On the second eigenvalue of hypergraphs*, Combinatorica, to appear.

[GG] O. Gabber and Z. Galil, *Explicit constructions of linear size superconcentrators*, Proc. 20th Annual Symposium on the Foundations of Computer Science, 1979, pp. 364–370.

[GHR] M.R. Garey, F.K. Hwang and G.W. Richards, *Asymptotic results for partial concentrators*, IEEE Trans. Comm. **36** (1988), 214–217.

[Ga] H. Garland, *p-adic curvature and the cohomology of discrete subgroups*, Ann. of Math. **97** (1973), 375–423

[Ge] S. Gelbart, «Automorphic Forms on Adele Groups», Princeton University Press, Princeton 1975.

[GJ] S. Gelbart, H. Jacquet, *A relation between automorphic representations of* $GL(2)$ *and* $GL(3)$, Ann. Scient. Éc. Norm. Sup. **11** (1978), 471–452.

[GGPS] I. Gelfand, Graev, I. Piatetski-Shapiro, «Representation Theory and Automorphic Functions», W.B. Saunders 1969.

[GVP] L. Gerritzen and N. Van der Put, «Schotcky Groups and Mumford Curves», Springer Lecture Notes in Mathematics **817** (1980).

[Go] D. Gorenstein, «Finite Simple Groups», Plenum Press, New York, 1982.

[Gd] R. Godement, *Domaines fondamentaux des groupes arithmetiques*, Séminaire Bourbaki 1962/3 exposé 257.

[GM] C.D. Godshil, B. Mohar, *Walk generating functions and spectral measures of infinite graphs*, Lin. Alg. and its App. **107** (1988), 191–206.

[Gr] E. Granirer, *Criteria for the compactness and discreteness of locally compact amenable groups*, Proc. Amer. Math. Soc. **40** (1973), 615–623.

[Gl] F.P. Greenleaf, «Invariant Means on Topological Groups», Van Nostrand, 1969.

[Gg] Y. Greenberg, *Ph. d. thesis*, Hebrew University, Jerusalem, in preperation.

[Gv] M. Gromov, *Groups of polynomial growth and expanding maps*, Publ. Math. IHES **53** (1981) 53–78.

[Gu] R.C. Gunning, «Lectures on Modular Forms», Ann. of Math. Studies, Princeton Univ. Press, Princeton 1962.

[GKS] R.M. Guralnick, W.M. Kantor and J. Saxl, *The probability of generating a classical group*, Comm. in Alg. **22** (1994), 1395–1402.

[H] W. Haemers, *Eigenvalue Techniques in Design and Graph Theory*, Mathematical Centre Tracts 121, North Holland, 1 Amsterdam 1979, 50–51.

[HW] G. Hardy and E. Wright, «An introduction to Number Theory», Claredon Press 1978.

[HV] P. de la Harpe and A. Valette, «La Propriete (T) de Kazhdan pour les Groupes Localement Compacts», Asterisque **175** Société Mathematique de France 1989.

[Ha] K. Hashimoto, *Zetafunctions of finite graphs and representations of p-adic groups*, Advanced Studies in Pure Math. **15** (1989), in: «Automorphic Forms and Geometry of Arithmetic Varieties», 211–280.

[He] I.N. Herstein, «Topics in Algebra», Blaisdell Publishing Co., New York 1964.

[Ho] A.J. Hoffman, *On eigenvalues and colorings of graphs*, in «Graph Theory and its Applications», (ed: B. Harris), Academic Press, 1970, pp. 79–91.

[HM] R.E. Howe and C.C. Moore, *Asymptotic properties of unitary representations*, J. Funct. Anal. **32** (1979), 72–96.

[HP] D.R. Hughes, E.C. Piper, «Design Theory» Cambridge University Press, Cambridge 1985.

[Hul] A. Hulanicki, *Means and Følner conditions on locally compact groups*, Studia Math. **27** (1966) 87–104.

[Hu] A. Hurwitz, *Über die Zahlentheorie der Quaternionen*, (1896) at «Hurwitz Mathematische Werke», Band II pp. 303–331, Birkhäuser, Basel 1933.

[Ih] Y. Ihara, *Discrete subgroups of $PL(2, k_p)$*, Proc. Symp. Pure Math. AMS **9** (1968), 272–278.

[Ih1] Y. Ihara, *On discrete subgroups of the two by two projective linear group over p-adic fields*, J. Math. Soc. of Japan **18** (1966), 219–235.

[Im] W. Imrich, *Explicit constructions of regular graphs without small cycles*, Combinatorica **4** (1984), 53–59.

[JL] H. Jacquet and R.P. Langlands, «Automorphic Forms on $GL(2)$», Springer Lect. Notes in Math. **114** (1970).

[JM] A. Jimbo and A. Maruoka, *Expanders obtained from affine transformations*, Combinatorica **7** (1987), 343–355.

[J] N. Jacobson, «Basic Algebra II», W.H. Freeman, San Francisco, 1980.

[Ja] J. Ja'Ja, *Time-space tradeoffs for some algebraic problems*, Proc. 12th Annual ACM Symposium on the Theory of Computing, 1980, pp. 339–350.

[K1] N. Kahale, *Better expansion for Ramanujan graphs*, 32-nd IEEE Symp. on Foundations of Computer Science FOCS(1991), 398–404.

[K2] N. Kahale, *On the second eigenvalue and linear expansion of regular graphs*, 33-rd IEEE Symp. on Foundations of Computer Science FOCS (1992) 296–303 (see also pp. 49–62 in [Fr3]).

[KR] N.J. Kalton and J.W. Roberts, *Uniformly exhaustive submeasures and nearly additive set functions*, Trans. AMS. **278** (1983), 803–816.

[Kan1] W.M. Kantor, *Some large trivalent graphs having small diameters*, Discrete Applied Math. **37** (1992) 353–357.

[Kan2] W.M. Kantor, *Some Cayley graphs of simple groups*, Discrete Appl. Math. **25** (1989), 99–104.

[Kan3] W.M. Kantor, *Some topics in asymptotic group theory*, in: »Groups, combinatorics and Geometry» (ed: M. Liebeck and J. Saxl) 403–412, Lond. Math. Soc. Lecture notes series 165, Cambridge Univ. Press 1992.

[KL] W.M. Kantor, A. Lubotzky, *The probability of generating a finite classical group*, Geometric Dedicata **36** (1990), 67–87.

[KP] R. Karp and N. Pippenger, *A time-randomness trade-off*, preprint.

[Kat1] N.M. Katz, *An estimate for character sums*, J. of the Amer. Math. Soc., **2** (1989), 197–200.

[Kat2] N.M. Katz, *Estimates for Soto-Andrade sums*, preprint.

[Ka] D.A. Kazhdan, *Connection of the dual space of a group with the structure of its closed subgroups*, Functional Anal. Appl. **1** (1967), 63–65.

[Ke1] H. Kesten, *Symmetric random walks on groups*, Trans. AMS **92** (1959), 336–354.

[Ke2] H. Kesten, *Full Banach mean values on countable groups*, Math. Scand. **7** (1959), 146–156.

[Kl1] M. Klawe, *Nonexistence of one-dimensional expanding graphs*, Proc. 22nd Annual Symposium on the Foundations of Computer Science, Nashville, TN, 1981.

[Kl2] M. Klawe, *Limitations on explicit constructions of expanding graphs*, SIAM J. Comput. **13** (1984), 156–166.

[Kn] M. Kneser, *Strong approximation*, in: Proceedings of Symposia in Pure Mathematics **9** (1967) 187–196.

[Kr] J. Krawezyk, *A remark on continuity of translation invariant functionals on $L_p(G)$ for compact Lie groups G*, Monatshefte für Mathematik **109** (1990), 63–67.

[LaRo] J. Lafferty and D. Rockmore, *Fast Fourier analysis for SL_2 over a finite field and related numerical experiments*, Expermental Mathematics **1** (1992) 45–139.

[LaRo2] J. Lafferty and D. Rockmore, *Numerical Investigation of the spectrum for certain families of Cayley graphs*, in [Fr3] pp. 63–73.

[La] S. Lang, «$SL_2(\mathbf{R})$», Springer-Verlag, New York, 1985.

[LM] T. Leighton, B. Maggs, *Expanders might be practical: Fast algorithms for routing around faults on multibutterflies*. In Proc. Foun. of Comp. Sci. (FOCS), (1989), 384–389.

[LT] T. Lengauer and R. Tarjan, *Asymptotically tight bounds on time-space trade-offs in a pebble game*, J. Assoc. Comput. Mach., **29** (1982), 1087–1130.

[Li] J. Li, *Kloosterman-Selberg zeta functions on complex hyperbolic spaces*, Amer. J. of Math. **113** (1991), 653–731.

[LPSS] J. Li, I. Piatetski-Shapiro and P. Sarnak, *Poincaré series for* $SO(n.1)$, Proc. Indian Acad. Sci. (Math. Sci.) **97** (1987), 231–237.

[Liw] W.W. Li, *Character sums and Abelian Ramanujan graphs*, J. of Number Th. **41** (1992), 199–217.

[LiS] M.W. Liebeck, A. Shaler, *The probability of generating a finite simple group*, Geom. Ded., to appear.

[LR] V. Losert and H. Rindel, *Almost invariant sets*, Bull. Lond. Math. Soc. **13** (1981), 145–148.

[Lu1] A. Lubotzky, *Group presentations, p-adic analytic groups and lattices in* $SL_2(\mathbb{C})$, Ann. of Math. **118** (1983), 115–130.

[Lu2] A. Lubotzky, *Lattices in rank one Lie groups over local fields*, Geometric and Functional Analysis (GAFA) **1** (1991), 405–431.

[Lu3] A. Lubotzky, *Subgroup growth and congruence subgroups*, Invent. Math., to appear.

[LPS1] A. Lubotzky, R. Phillips and P. Sarnak, *Ramanujan conjectures and explicit construction of expanders*, Proc. Symp. on Theo. of Comp. Sci. (STOC), **86** (1986), 240–246.

[LPS2] A. Lubotzky, R. Phillips and P. Sarnak, *Ramanujan graphs*, Combinatorica **8** (1988), 261–277.

[LPS3] A. Lubotzky, R. Phillips and P. Sarnak, *Hecke operators and distributing points on* S^2, I, Comm. Pure and Applied Math. **39** (1986), S149–186.

[LPS4] A. Lubotzky, R. Phillips and P. Sarnak, *Hecke operators and distributing points on* S^2*, II*, Comm. Pure and Applied Math. **40** (1987) 401–420.

[LS] A. Lubotzky, A. Shalev, *On some* Λ*-analytic pro-p groups*, Israel Journal of Math. **85** (1994), 307–339.

[LW] A. Lubotzky, B. Weiss, *Groups and expanders*, in: «Expanding graphs» 95–109, DIMACS series Vol. 10, American Math. Soc. 1993, (Ed: J. Friedman).

[LZ] A. Lubotzky and R.J. Zimmer, *Variants of Kazhdan's property for subgroups of semi-simple groups*, Israel J. Math. **66** (1989), 289–299

[Ma] W. Magnus, *Residually finite groups*, Bull. A.M.S. **75** (1969), 305–316.

[M1] G.A. Margulis, *Explicit constructions of concentrators*, Probl. of Inform. Transm. **10** (1975), 325–332.

[M2] G.A. Margulis, *Some remarks on invariant means*, Monatshefte fur Mathematik **90** (1980), 233–235.

[M3] G.A. Margulis, *Explicit constructions of graphs without short cycles and low density codes*, Combinatorica **2** (1982), 71–78.

[M4] G.A. Margulis, *Arithmetic groups and graphs without short cycles*, 6th International Symp. on Information Theory, Tashkent 1984, Abstracts, Vol. **1**, pp. 123–125 (in Russian).

[M5] G.A. Margulis, *Some new constructions of low-density paritycheck codes*, 3rd Internat. Seminar on Information Theory, convolution codes and multi-user communication, Sochi 1987, pp. 275–279 (in Russian).

[M6] G.A. Margulis, *Explicit group theoretic constructions of combinatorial schemes and their applications for the construction of expanders and concentrators*, J. of Problems of Information Transmission, **24** (1988), 39–46.

[M7] G.A. Margulis, «*Discrete Subgroups of Semisimple Lie Graphs*», Springer-Verlag 1991.

[MJ] G. Masson and B. Jordan Jr., *Generalized multi-stage connection networks*, Networks, **2** (1972), 191–209.

[MK] B. McKay, *The expected eigenvalue distribution of a large regular graph*, Lin. Alg. Appl. **40** (1981), 203–216.

[Me] J.F. Mestre, *Les méthode des graphes; Examples et applications*, in «Proceedings, International conference on class Numbers and Fundamental Units of Algebraic Number Fields, June 24–26, 1986, Katata, Japan», pp. 217–242.

[Moh] B. Mohar, *Isoperimetric numbers of graphs*, Journal of Combinatorical Theory B **47** (1989), 274–291.

[Mor1] M. Morgenstern, *Existence and explicit construction of $q+1$ regular Ramanujan graphs for every prime power q*, J. of Comb. Th. B, to appear.

[Mor2] M. Morgenstern, *Ramanujan diagrams*, SIAM J. of Disc. Math, to appear.

[Mor3] M. Morgenstern, *Explicit construction of natural bounded concentrators*, 32nd IEEE Symp. on the Foundations of Computer Science (FOCS) 1991, 392–397. (An extended version to appear in Combinatorica.)

[Mz] S. Mozes, *A zero entropy, mixing of all orders tiling systems*, Contemporary Math. **135** (1992), 319–325.

[Mo] G.D. Mostow, *Discrete subgroups of Lie groups*, in: «Lie Theories and their Applications», Queen's Papers in Pure and Appl. Math. No. 48 (ed.: A. J. Coleman and P. Rubenboim) pp. 65–153, Queen's University, Kingston, Ontario 1978.

[MT] G.D. Mostow and T. Tamagawa, *On the compactness of arithmetically defined homogeneous spaces*, Ann. of Math. **76** (1962), 440–463.

[Na] I. Namioka, *Fϕlner condition for amenable semigroups*, Math. Scand. **15** (1964), 18–28.

[Ni] A. Nilli, *On the second eigenvalue of a graph*, Discrete Math. **91** (1991), 207–210.

[Of] Ju.P. Ofman, *A universal automaton*, Trans. Moscow Math. Soc. **14** (1965), 200–215.

[Og] A. Ogg, «Modular Forms and Dirichlet Series», Benjamin Inc. New York, 1969.

[Or] Oystein Ore, «Theory of Graphs», Amer. Math. Soc. Coll. Publ. Vol. 38, AMS Providence, RI, 1962.

[P] A.L.T. Paterson, «Amenability», Mathematical Surveys and Monographs Vol. 29, Amer. Math. Soc., Providence, R.I. 1988.

[PT] W. Paul and R. Tarjan, *Time-space trade-offs in a pebble game*, Acta Inform. **10** (1978), 111–115.

[PTC] W. Paul, R. Tarjan and J. Celoni, *Space bounds for a game on graphs*, Math. Systems Theory **10** (1979), 239–251.

[Pa] I. Pays, *Graphs arithmetic*, preprint.

[Pie] J.-P. Pier, «Amenable Locally Compact Groups», John Wiley & Sons, New York, 1984.

[Pi] M. Pinsker, *On the complexity of a concentrator*, in 7th International Teletraffic Conference, Stockholm, June 1973, 318/1–318/4.

[P1] N. Pippenger, *Superconcentrators*, SIAM J. Comput. **6** (1977), 298–304.

[P2] N. Pippenger, *Generalized connectors*, J. Comput. System Sci. **7** (1978), 510–514.

[P3] N. Pippenger, *Pebbling with an auxiliary pushdown*, J. Comput. System Sci. **23** (1981), 151–165.

[P4] N. Pippenger, *Sorting and selecting in rounds*, SIAM J. Comput. **16** (1987), 1032–1038.

[P5] N. Pippenger, *Communication Networks*, in: «Handbook of Theoretical Computer Science», Edited by J. Van Leeuwen, Elsevier Science Publications, B.V., (1990), 806–833.

[Piz] A. Pizer, *Ramanujan graphs and Heche operator*, Bull. AMS **23** (1990), 127–137.

[PR] V. Platonov, Rapinchuk, *Abstract characterzation of arithmetic groups with the congruence subgroup property*, Dokl. Math. USSR **319** (1991) 1322–1327.

[Pr1] G. Prasad, *Elementary proof of a theorem of Tits and a theorem of Bruhat-Tits*, Bull. Soc. Math. Fr. **110** (1982), 197–202.

[Pr2] G. Prasad, *Strong approximation for semi-simple groups over function fields*, Ann. of Math. **105** (1977), 553–572.

[Ra1] M.S. Raghunathan, «Discrete Subgroups of Lie Groups», Springer-Verlag, New York, 1968.

[Ra2] M.S. Raghunathan, *On the congruence subgroup problem*, Publ. Math. IHES **46** (1976), 107–161.

[Rn] B. Randl, *Small eigenvalues of the Laplace operator on compact Riemann surface*, Bull. A.M.S. **80** (1974), 996–1000.

[Re] I. Reiner, «Maximal Orders», Academic Press, (1975).

[Rn] Y. Roichman, *Caley graphs of the symmetric groups*, Hebrew University thesis 1994 (in Hebrew).

[Ro1] J. Rosenblatt, *Uniqueness of invariant means for measure preserving transformations*, Trans. AMS **265** (1981), 623–636.

[Ro2] J. Rosenblatt, *Translation-invariant linear forms on $L_p(G)$*, Proc. A.M.S. **94** (1985), 226–228.

[Ro3] J. Rosenblatt, *Automatic continuity is equivalent to uniqueness of invariant means*, Ill. J. Math. **35** (1991), 339–348.

[Ru1] W. Rudin, *Invariant means on L^∞*, Studia Math. **44** (1972), 219–227.

[Ru2] W. Rudin, «Functional Analysis», McGraw-Hill, New York, 1973.

[Sa1] P. Sarnak, *The arithmetic and geometry of some hyperbolic three manifolds*, Acta Math. **151** (1983), 253–295.

[Sa2] P. Sarnak, «Some Applications of Modular Forms», Cambridge Tracts in Mathematics **99**, Cambridge University Press 1990.

[Sa3] P. Sarnak, *Number theoretic graphs*, unpublished preprint 1988.

[Sat] I. Satake, *Spherical functions and Ramanujan conjecture*, in: «Algebraic Groups and Discontinuous Subgroups», Proc. Symp. in Pure Math. Vol. IX, (1967), 258–264.

[Sc 1] K. Schmidt, *Asymptomatically invariant sequences and an action of $SL_2(\mathbf{Z})$ on the 2-sphere*, Israel J. of Math. **37** (1980), 193–208.

[Sc 2] K. Schmidt, *Amenability, Kazhdan's property T, strong ergodicity and invariant means for ergodic group actions*, Erg. Th. and Dyn. Syst. **1** (1981), 223–236.

[Se] A. Selberg, *On the estimation of fourier coefficients of modular forms*, Proc. Symp. Pure Math. **VIII** (1965), 1–15.

[S1] J.J. Serre, «Trees», Springer-Verlag, Berlin-Heidelberg-New York, 1980.

[S2] J.P. Serre, *Le problem des groupes de congruence pour SL_2*, Ann. of Math. **92** (1970), 489–527.

[S3] J.P. Serre, «Linear Representations of Finite Groups», Springer-Verlag, New York, 1977.

[Sl] N.J.A. Sloane, *Encrypting by random rotations*, Cryptography, Lecture Notes in Computer Science **149**, pp. 71–128 (ed: T. Beth).

[SW] P.M. Soardi, W. Woess, *Amenability, unimodularity and the spectral radius of random walks on infinite graphs*, Math. Z. **205** (1990), 471–486.

[Su] D. Sullivan, *For $n > 3$ there is only one finitely additive rotationally invariant measure on the n-sphere on all Lebesgue measurable sets*, Bull. AMS **1** (1981), 121–123.

[Sd] T. Sunada, *Fundamental groups and Laplacians*, in: Lecture Notes in Mathematics 1339, (T. Sunada editor), Springer Varlag, (1987), 248–277.

[T] K. Takeuchi, *Arithmetic triangle groups*, J. Math. Soc. Japan **29** (1977), 91–106.

[Ta] R.M. Tanner, *Explicit concentrators from generalized N-gons*, SIAM J. Alg. Discr. Meth., **5** (1984), 287–294.

[Te1] A. Terras, «Fourier Analysis on Finite Groups and Applications», UCDS Notes.

[Te2] A. Terras, *Eigenvalue problem related to finite analogues of upper half planes*, in: «Forty More Years of Ramifications: Spectral Asymptotics and its Applications», (ed: S.A. Fulling and F.J. Narcowicl), Math. Dept., Texas A & M, college Station, TX, Discourses in Math. **1** (1991).

[Te3] A. Terras, *Are finite upper half plane graphs Ramanujan?*, in [Fr3] pp. 125–142.

[Ti] J. Tits, *Free subgroups in linear groups*, J. of Algebra **20** (1972), 250–270.

[To] M. Tompa, *Time-space tradeoffs for computing functions, using connectivity properties of their circuit*, J. Comput. System Sci. **20** (1980), 118–132.

[Up] E. Upfal, *An $O(\log n)$ deterministic packet routing scheme*. In: Proc. Symp. on Theo. of Comp. Sci. (STOC), (1989), 241–250.

[Va] L. Valiant, *Graph theoretic properties in computational complexity*, J. Comput. System Sci. **13** (1976), 278–285.

[Vs] L.N. Vaserstein, *Groups having the property* (T), Funct. Anal. and its Appl. **2** (1968), 174.

[Vi1] M.F. Vignéras, «Arithmétique des Algébres de Quaternions», Lecture Notes in Mathematics **800**, Springer-Verlag, 1980.

[Vi2] M.F. Vignéras, *Quelques remarques sur la conjecture* $\lambda_1 \geq \frac{1}{4}$, Seminaire de theory des nombres, Delange-Pisot-Poiton, Paris (1981/2), Progr. Math. **38**, Birkhäuser, (1983), 321–343).

[Wa] S. Wagon, *The Banach-Tarski Paradox, «Encyclopedia of Mathematics and its Applications»*, Volume 24, Cambridge University Press, Cambridge, 1985.

[Wan1] S.P. Wang, *The dual space of semi-simple Lie groups*, Amer. J. Math. **91** (1969), 921–937.

[Wan2] S.P. Wang, *On isolated points in the dual spaces of locally compact groups*, Math. Ann. **218** (1975), 19–34.

[War] F.W. Warner, *Foundation of Differential Manifolds and Lie Groups*, Scott-Foresman and Company, Glenview, Illinois, 1971.

[Wt] Y. Watatani, *Property* (T) *of Kazhdan implies property* (FA) *of Serre*, Math. Japon. **27** (1981), 97–103.

[We] A. Weiss, *Girths of bipartite sextet graphs*, Combinatorica **4** (1984) 241–245.

[Wi] F. Williams, «Lectures on the Spectrum of $L^2(\Gamma \setminus G)$», Pitman Research Notes in Math. no. 242, Longman Sci. and Tech. 1991.

[Zi1] R.J. Zimmer, «Ergodic Theory and Semi-Simple Groups», Birkhäuser, Boston, 1984.

[Zi2] R.J. Zimmer, *Kazhdan groups acting on compact manifolds*, Invent. Math. **75** (1984), 425–436.